Sick Building Syndrome

Sick Building Syndrome

Concepts, issues and practice

Edited by

Jack Rostron

E & FN SPON
An Imprint of Chapman & Hall

London · Weinheim · New York · Tokyo · Melbourne · Madras

Published by E & FN Spon, an imprint of Chapman & Hall, 2–6 Boundary Row, London SE1 8HN, UK

Chapman & Hall, 2–6 Boundary Row, London SE1 8HN, UK

Chapman & Hall GmbH, Pappelallee 3, 69469 Weinheim, Germany

Chapman & Hall USA, 115 Fifth Avenue, New York, NY 10003, USA

Chapman & Hall Japan, ITP-Japan, Kyowa Building, 3F, 2-2-1 Hirakawacho, Chiyoda-ku, Tokyo 102, Japan

Chapman & Hall Australia, 102 Dodds Street, South Melbourne, Victoria 3205, Australia

Chapman & Hall India, R. Seshadri, 32 Second Main Road, CIT East, Madras 600 035, India

First edition 1997

© 1997 E & FN Spon

'Little boxes' by Malvina Reynolds
© 1962 Schroder Music Co., USA
Assigned to TRO ESSEX MUSIC LTD, London
All rights reserved. International copyright secured.
Used by permission.

Typeset in 10/12pt Palatino by Columns Design Ltd, Reading
Printed in Great Britain by TJ International, Padstow, Cornwall

ISBN 0 419 21530 1

A catalogue record for this book is available from the British Library

∞ Printed on permanent acid-free text paper, manufactured in accordance with ANSI/NISO Z39.48-1992 and ANSI/NISO Z39.48-1984 (Permanence of Paper).

Contents

Foreword

Little boxes on the hillside
Little boxes made of ticky-tacky
Little boxes, little boxes
And they all look just the same
(Malvina Reynolds)

The post-war rush to rebuild produced many eyesores and few master-pieces. It seemed that the human element – the occupier – was at the bottom of the list in a world seeking to accommodate ever changing technology and corporate ambition to maximize profitability. As we approach the millennium let us hope that the human element will rise to the top – its time is certainly now. This means creating working environments which must not only meet the needs of the computer and finance director but also their servants: mankind. As chartered surveyors we seek to meet our clients' objectives by acquiring, developing, letting, building and managing commercial property. Our clients in turn seek to meet their customers' needs in whatever business they are focused and, yes, to make a profit.

At the end of the day it is encouraging to think that the needs of all can be met by producing better buildings and better working environments. A comprehensive understanding of the impact of design, layout and environmental control on occupiers is long overdue. It is heartening that the human element is now being examined to see how he or she works best and, importantly, what offends or upsets. Fashion will no doubt dictate from time to time that little boxes are built and perhaps they will all look just the same; with care and attention, however, those working within them and those owning them may profit from the wisdom and endeavour brought together by Jack Rostron in this anthology.

Ian Dodwell
Weatherall Green & Smith
Norfolk House
31 St James's Square
London SW1Y 4JR
May 1996

Preface

This book attempts to expose the reader to the enigma of sick building syndrome. The reader is introduced to the many facets of the syndrome from potential causes, through the consequences and to possible remedies. The specialist contributors take the reader on a journey through the medical, legal, architectural and management issues which the syndrome impinges on. The book is written in a way to offer both theoretical and practical guidance.

It does not claim to be exhaustive on what it attempts. It does not claim to do everything. But it does aim to give the reader an overall picture about the syndrome. I hope it will give sufficient knowledge to readers to help them reduce the impact of this sometimes devastating disease. If it does achieve this objective, in whatever degree, I shall be more than pleased.

There are many to whom an insight into the syndrome will be of interest: physicians, lawyers, architects, engineers, personnel officers, developers, etc. If they find the book of some assistance, our efforts will have been worth while.

Jack Rostron   May 1996
Liverpool John Moores University
Clarence Street
Liverpool L3 5UG

Acknowledgements

I am indebted to many people, both directly and indirectly. My particular thanks are to those who contributed directly to the book: Judge John Clark of the Northern Circuit; Dr Hana Drahonovska of the National Institute of Public Health, Prague; Dr Keith Eaton of Princess Margaret Hospital; Adrian Leaman of the Institute for Advanced Architectural Studies, University of York; William Bordass of Bordass Associates, London; Dr Chris Baldry, Peter Bain and Philip Taylor of Strathclyde University; Alison Raynor of John Moores University; Nick McCallen and Rob Davies. Without their contribution this book would not have been completed.

I am particularly indebted to Dr Keith Eaton for both contributing a chapter to the book and reviewing initial drafts and to Charles Danecki for making suggested improvements.

Thanks must also go to Ian Dodwell and Paul Gardner of Weatherall, Green & Smith for supporting my initial research into sick building syndrome.

The staff of the School of the Built Environment, Liverpool John Moores University have been unstinting in their support in many ways.

I am grateful to Johnson Control Systems for permission to reproduce several drawings.

The editorial staff of E & F N Spon have been supportive throughout in bringing this publication to completion: Rachael Wilkie, Mike Doggwiler and Tim Robinson.

Contributors

Peter Bain, Lecturer, BA, MSc, Department of Human Resources Management, Strathclyde University, Glasgow.

Dr Chris Baldry, Senior Lecturer, PhD, MSc, Department of Human Resources Management, Strathclyde University, Glasgow.

Dr Bill Bordass, PhD, Principal William Bordass Associates, London.

Judge John Clark, LLB, Deputy District Judge of the Northern Circuit and Partner Carter Hodge, Solicitors, Southport.

Robert Davies, BSc, MScEng, Consultant Building Services Engineer, Liverpool.

Dr Hana Drahonovska, MD, PhD, Head of the National Laboratory for Indoor Environment, National Institute of Public Health, Prague.

Dr Keith Eaton, LRCP&SE, LFRP&SG, Consultant Physician, Princess Margaret Hospital, Windsor.

Adrian Leaman, BA, Director of Building Use Studies Ltd, Director of Research at the Institute for Advanced Architectural Studies, University of York.

Nick McCallen, MSc, Consulting Software Engineer, Southport.

Alison J. Rayner, BSc, MCIOB, Senior Lecturer in Environmental Science, School of the Built Environment, Liverpool John Moores University.

Vyla Rollins, BA, MPhil, C, Psych AFBPS, Organizational Psychologist, Kinsley Lord, London.

Jack Rostron, MA, ARICS, MRTPI, sometime Adviser to the World Health Organization and BPS Research Fellow, School of the Built Environment, Liverpool John Moores University.

Gill-Helen Swift, BSc, MSc, BPS, Environmental Psychology Consultant, Procord Ltd, Portsmouth.

Philip Taylor, MA, M.Phil, Research Fellow, Department of Human Resources Management, Strathclyde University, Glasgow.

CHAPTER 1

Introduction

Jack Rostron

It is clearly important that buildings should provide a healthy, safe and comfortable environment for their occupants. A relatively recent phenomenon, sick building syndrome, SBS [1] has been acknowledged as a recognizable disease by the World Health Organization [2]. Since its recognition in 1986, increasing concern has been directed towards identifying a cause and eliminating it from occupied buildings or those at the design stage [3–8].

It has been estimated [2] that up to 30% of refurbished buildings and an unknown but significant number of new buildings may suffer from SBS. The consequences of the syndrome are, *inter alia*, increased absenteeism, reduced work performance and possibly even building closure. While it is not life threatening or disabling it is clearly important to those affected by it.

The effect of SBS on individuals is a group of symptoms which are experienced specifically at work. The typical symptoms are headache; loss of concentration; itchy, runny or stuffy nose; itching, watering or dry eyes; dry skin; lethargy; and dryness or irritation of the throat. Such symptoms exist in the general population, but are distinguished by a higher incidence, as a group, in some buildings than others.

SCOPE AND PURPOSE

It is against the background of this increasing concern, from the point of view of both the health of individual workers, and the productivity of organizations employing them, that this book was prepared. By inviting the views of recognized specialists in diverse fields it attempts to offer both practical guidance and up-to-date discussion on certain aspects.

The contributions are aimed at a diverse readership, ranging from those looking for an immediate practical 'solution' to students and academic researchers interested in exploring the latest thinking on concepts and techniques.

Chapter 2 offers a detailed overview of the possible causes of SBS and makes recommendations for improving the internal work environment. It deals in synoptic form with heating and ventilation systems, environmental science aspects, control systems and the design and maintenance process.

Chapter 3 offers an examination of the concept of light and lighting and its relationship to the health of building occupants. It describes the physical nature of light and how it can cause SBS. The importance of natural and artificial light is explored. The health consequences of inappropriate lighting conditions is explained and means of designing lighting systems to reduce the incidence of SBS are suggested.

The medical aspects of SBS are discussed in Chapter 4. The syndrome as a medical concept is evaluated and thoughts on its investigation and management from the clinical viewpoint are explored.

Chapter 5 examines the psychological issues associated with the syndrome. The different types of workplace illness are discussed and suggestions for a multidisciplined approach to the subject are postulated.

Maintenance of the working environment has been recognized as an important means of reducing the syndrome's incidence rate. Chapter 6 describes in a practical way the principal procedures necessary to implement an effective plant maintenance programme and cleaning regime. It is written in a simple and easily understandable style suitable for facilities managers who have not acquired an in-depth knowledge of the engineering issues involved.

The importance of human resource management in reducing the impact of SBS is comprehensively discussed in Chapter 7. The latest research findings and personnel management concepts are evaluated in terms of restructuring office work, the relationship between technology and people, personnel policies to help cope with the syndrome, the role of the occupational health service, worker participation and the physical environment.

The legal implications of SBS are increasingly becoming recognized. Chapter 8 explains the law both in the UK and the European Community. Both statute and common law aspects are stated and their significance is explained.

Since the first studies of SBS in the mid-1980s, the health scares over legionnaires' disease and international efforts to reduce energy consumption in buildings, among other factors, have helped to encourage research effort into the human, social and environmental aspects of buildings. Chapter 9 explores some of the recent concepts affecting thinking about building management and design. Their implications are examined, especially for strategic thinking about buildings and their occupants.

A practical and 'simple' to use procedure to identify and rectify the existence of SBS in buildings is described in Chapter 10. A synopsis of the main promoters of the syndrome is offered and each known cause is reduced to a checklist which facilitates 'easy' diagnosis and an indication of the remedies.

The importance of advanced computational techniques in diagnosing and responding to SBS is reviewed in an appendix [9] which describes an expert system SBARS v2.1 [10] that offers a process for assessing the existence of sick building syndrome in existing and new buildings. A rapid and cost-effective process for rectifying existing buildings or eliminating the known promoters of SBS at the design stage of new buildings is discussed.

Despite extensive international research efforts, there is still disagreement about the cause of SBS. While there is common agreement about the deleterious effect the syndrome has on people, the search for a cure is being pursued.

THE FUTURE

The future concerning SBS is mixed. On the one hand some believe that continuing organizational and technological change which affect the way employees work may be adding further troublesome ingredients to an already unpleasant cocktail of factors adversely affecting employee morale and well-being. The government's decision to shift responsibility for sick pay to the employer is bound to create tighter control over persistent absentees, such as those suffering from the syndrome, who may feel it necessary to return to work, although the work environment could perhaps have been the cause of their illness in the first place.

Another equally valid viewpoint is that the current weight of international scientific research effort [3, 4, 5, 10], may yield the specific cause of the syndrome. It may then be possible to affect a 'cure' and eradicate the serious consequences which SBS currently brings.

It is hoped that in some way the various contributions contained in this book will assist in ameliorating some of the problems which the syndrome has brought to the world of work.

REFERENCES

1. Sykes, J.M. (1988) *Sick Building Syndrome: A Review*, Health and Safety Executive, Specialist Inspector Reports No. 10.
2. Akimenko, V.V., Anderson, I., Lebovitze, M.D. and Lindvall, T. (1986) The sick building syndrome, *Indoor Air*, 6, Swedish Council for Building Research, Stockholm, pp. 87–97.
3. Leinster, P., Raw, G., Thomson, N., Leaman, A. and Whitehead, C. (1990) *A Modular Longitudinal Approach to the Investigation of Sick Building Syndrome*, Building Research Establishment, Garston.

4. Whorton, M.D. (1987) Investigation and work up of tight building syndrome, *Journal of Occupational Medicine*, **29**.
5. Wilson, S. and Hedge, H. (1987) *The Office Environment Survey: A Study of Building Sickness*, Building Use Studies Ltd, London.
6. Sterling, T.D. and Sterling, E.M. (1987) Environmental tobacco smoke and indoor air quality in modern office environments, *Journal of Occupational Medicine*, **29**.
7. Hedge, A. and Collis, M.D. (1987) Do negative ions affect human mood and performance?, *Annals of Occupational Hygiene*, **31** (3), 285–90.
8. Valbjern, O. and Skov, P. (1987) Influence of indoor climate on the sick building syndrome, *Proceedings of the 4th International Conference on Indoor Air Quality and Climate*, Vol. 2, Institute for Water, Soil and Air Hygiene, Berlin.
9. Rostron, J. (1994) Sick building syndrome, *Public Service and Local Government*, May.
10. SBARS v2.1 (1994) *Sick Building Assessment and Rehabilitation System*, Environmental Management and Intelligent Information Systems Ltd, EMIIS Ltd, Salford.

Overview of the possible causes of SBS and recommendations for improving the internal environment

Alison J. Rayner

Included in this chapter are some of the possible causes of SBS and possible ways of eradicating these problems. Problem areas have been identified from research carried out extensively on a worldwide basis. Much research has been carried out to try to prove conclusively the cause of SBS. This unfortunately has not as yet been achieved, with many of the previously held ideas being disproved. Heating and ventilation systems, air quality, noise levels, humidity and temperature will be investigated for their contribution to SBS.

COMMON FEATURES OF SBS

The range of symptoms and their prevalence depends on various factors. Studies in the UK show a number of common factors [1]:

1. Symptoms are most common in air conditioned buildings but can also occur in naturally ventilated buildings.
2. Clerical staff are more likely to suffer than managerial staff, and more complaints arise in the public than the private sector.
3. People with most symptoms have least perceived control over their environment.
4. Symptoms are more frequent in the afternoon than the morning.

The World Health Organization [2] identifies a number of features that are common to sick buildings [1]:

1. They often have forced ventilation (the WHO does not specifically refer to air conditioning, even though it falls into this category).
2. They are often of light construction.
3. Indoor surfaces are often covered in textiles.
4. They are energy efficient, kept relatively warm and have a homogeneous thermal environment.
5. They are airtight, i.e. windows cannot be opened.

All of these aspects need to be looked at when trying to ascertain a cause and other possible contributory factors.

HEATING AND VENTILATING SYSTEMS

Mechanical ventilation of buildings differs from natural ventilation in a number of ways:

• With natural ventilation occupiers have a choice, i.e. they can open the window when it is warm and shut it when it is cold. It is entirely down to the individual's comfort requirements. This is not the case in mechanically 'sealed' systems as the control is fixed throughout the building.
• The air supply in mechanical ventilation systems can be varied so that a higher proportion of recirculated air to fresh air is passed around the building.
• Any mechanical system can be subject to design faults and poor installation. Components may fail and have a devastating impact on the performance of the system.

Fresh air is required in an air conditioning system to supply air for respiration and to dilute CO_2, odours, smoke and other contaminants. Ventilation is required to maintain personal comfort, i.e. temperature control. Several standards have been set for ventilation and fresh air rates to offices, usually based on the air required to dilute cigarette smoke or body odours. As these standards are constantly revised and can vary considerably in different situations it is difficult to state an acceptable level.

The need to ventilate mechanically has increased with the increase in building plan size. With a small plan area with a set number of offices it is easy to ventilate naturally using opening windows (see Fig. 2.1). Each office and corridor has its own window, therefore occupants can choose their environment. The growth in large building plans with open plan office arrangements has made this impossible, especially when a building is 'sandwiched' between two others (see Fig. 2.2). All the evidence

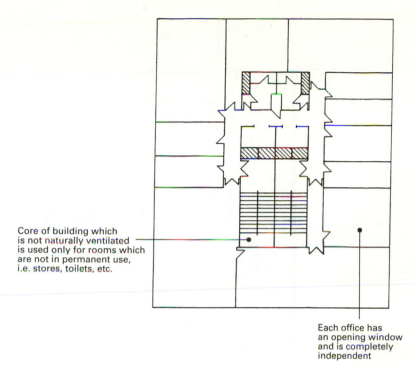

Core of building which
is not naturally ventilated
is used only for rooms which
are not in permanent use,
i.e. stores, toilets, etc.

Each office has
an opening window
and is completely
independent

Fig. 2.1 Traditional office layout plan.

suggests that inadequate fresh air is a contributory factor in SBS, although nothing has yet been proved.

Relative humidity

Low relative humidity can lead to an increase in the incidence of respiratory infection [1]. The *CIBSE Guide* [3] recommends that humidity should be maintained between 40 and 70%. This is a broad band, and testing could only prove that low humidity was the cause of SBS if tests were carried out on buildings with the same humidity. The reasons given as to why humidity can increase the symptoms of SBS are as follows:

- Humidity may affect the survival of bacteria and viruses; air-borne micro-organisms are less likely to survive in relative humidities of the order of 50%. However, it is also true to say that very high humidity causes dampness, which can encourage the growth of micro-organisms.
- Higher humidities encourage the agglomeration of air-borne particles, and larger particles are believed to be less likely to cause infection than smaller ones.

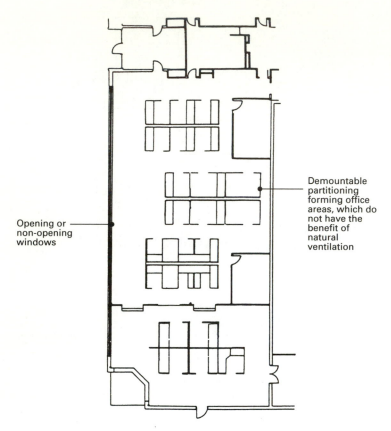

Opening or
non-opening
windows

Demountable
partitioning
forming office
areas, which do
not have the
benefit of
natural
ventilation

Fig. 2.2 Typical open plan office layout.

- Dry air may produce micro-fissures in the upper respiratory tract which may act as landing sites for infection.
- Increased mucus flow favours rejection of micro-organisms. (However there is still insufficient evidence at this time to prove this point.)

Low humidity has been proved to not be the major causes of SBS in buildings; however, it is known to cause some of the symptoms. Erythema (skin rash) may also be caused by low humidity and research has proved that increasing humidity from 30–35% to 40–45% can go some way to alleviating this problem.

As has been stated previously, various standards have been set for the optimum comfort of building occupants, the most widely accepted being that derived by Fanger and used as the basis for ISO 7730-1984 [7]. Like most standards, ISO 7730-1984 sets an optimum temperature range (air, radiant and radiant symmetry) for people at different metabolic rates and wearing different clothing. Recommended comfort requirements from ISO 7730-1984 are:

1. operative temperature 20–24°C (22 +/-2°C);
2. vertical air temperature difference 1.1 m and 0.1 m (head and ankle height) less than 3°C;
3. floor surface temperature 19–26°C (29°C with floor heating systems);
4. mean air velocity less than 0.15 m/s;
5. radiant temperature asymmetry (due to windows, etc. less than 10°C;
6. radiant temperature asymmetry from a warm ceiling less than 5°C.

The standard is based on the predicted mean vote (PMV) and the predicted percentage of dissatisfied (PPD), which predicts conditions which are most satisfactory to most people for most of the time. These percentages are obtained from people working within the environment, expressing their ideals.

Ideal conditions can vary from population to population, and from person to person within a population, therefore we can make the assumption that there is no ideal thermal environment; what is ideal for one person is not for another.Therefore the more people that work in any particular environment the larger the number of people likely to be dissatisfied.

The standards set in ISO 7730-1984 are complex and difficult to achieve. In a closed environment if the mechanical system fails or is not working at 100%, there is little or nothing the occupants can do. Stuffiness within the environment is also a complaint. Stuffiness can be caused by a lack of adequate air velocity. The mechanical system should allow for this; but what happens when it fails? The use of free-standing fans may help the problem but the majority of employers who either own or lease a property with a mechanical system would be loath to supply these.

There is some evidence that SBS may be linked to the use of humidifiers, being caused by micro-organisms in the humidifier. There may be other sites where micro-organisms can breed, for example in furnishings. Thus the presence of micro-organisms may be significant. However, an infection is unlikely to be the cause since symptoms disappear quickly when away from the workplace. An allergy would fit the pattern better, but allergies do not affect people in such numbers and in differing ways.

The contribution of artificial heating and ventilation systems to SBS

The World Health Organization says the cause of SBS is unknown, but researchers at Strathclyde University [8] say that evidence points to construction changes after the 1970s' fuel crisis, which led to sealed buildings with temperature, ventilation and lighting centrally controlled, often by computer. This leads us to believe that air conditioning could be the major cause of the problem, and much research has been carried out to test this belief.

America's Environmental Protection Agency [9] determined that 20–35% of office workers suffer from poor air quality at work, which may have some contribution to SBS, or certainly create some of the symptoms. However, it would be fair to say that poor air quality is not the sole cause. In some buildings with very poor air quality the symptoms of SBS among workers are not as bad as those in buildings with better quality air.

So what does cause the problem? America's National Institute of Occupational Safety and Health has carried out extensive research on SBS [9], and has apportioned 50% of the blame on inadequate ventilation. Poor air quality created by microbes, volatile substances, particles from fabrics and internal or external contaminants accounts for approximately 25%, the rest remaining undetected. The other problem with using this type of statistical information is that it can be unreliable, and can only be proved conclusively when improvements have been implemented, the results monitored and the end result is satisfactory.

Research carried out by Mrs Sheena Wilson of Building Use Studies, Dr Alan Hedge of Cornell University in New York, Dr Peter Sherwood-Burge of Solihull Hospital and Dr Jon Harris-Bass on 46 buildings in Britain found that the buildings rated the most healthy were those with opening windows, while the sickest were those that had sealed windows [9]. This may well have contributed to the assumption that the opening of windows is a solution, and has led to the widespread belief that the cause of SBS is mechanical ventilation systems. One would also have to assume that it is the air quality of artificially air conditioned buildings that is the root cause of the problem simply because in buildings with opening windows outside pollution will have an impact as the air is not filtered. These assumptions reinforce the theory that SBS only occurs in post-1960s' buildings that are mainly open plan and either had an artificial ventilation system installed at the time of construction or have had one installed subsequently. One of the main causes of serious disease in this country in earlier days was cross-infection arising from poorly ventilated domestic properties; recirculating air through a building could also lead to cross-infection of office workers.

However, research carried out by the team proved that the actual air quality in such buildings was not poor. The air was less stale than in naturally ventilated buildings. This study also found that people tended to blame the 'dryness of the air for their discomfort. Again, however, humidity levels were checked and sick buildings proved to be no worse than their naturally ventilated counterparts. Smoking has also been blamed, but this can be discounted as a major cause as people who suffer from SBS and smoke tend to blame their symptoms on smoking; so why do people not suffer from SBS in their own homes, and why do all those who work in offices with artificial ventilation and smoke, not suffer from the symptoms of SBS?

The other area that causes concern is that SBS could in some way be linked to the materials used in the fabric of the building construction interacting with other materials and the services systems to create unique environmental problems within buildings [1]. The remedy for this would be to test all building components in closed environments with every other material that could be encountered. This solution would be nigh on impossible, and therefore assumptions have to be made by those specifying materials and designing services systems. All of the research carried out to prove whether the main cause of SBS in buildings is artificial ventilation systems, and to a large extent the earlier research, has proved this to be a major cause. However, more recent research carried out by the BSRIA [10], which carried out extensive tests has shown that the most sick building was air conditioned – but so was the least sick. The naturally ventilated office scored somewhere in the middle. The BSRIA concluded that there was nothing inherently unhealthy about air conditioned offices [11]. It reported that 'air conditioned buildings, when well designed and maintained, can be associated with very low levels of symptoms attributable to SBS'.

One of the main reasons why air conditioning systems have been associated so strongly with SBS in the past is that the majority of buildings studied have been open plan offices with air conditioning. For this reason it was felt that there must be a link. The buildings are extremely complex, with an enormous number of components, and an absolute answer is very hard to find. One of the other areas that causes concern is the cleanliness of the air conditioning system when installing. Dirty ducts that are not cleaned prior to commissioning will cause the dirt to be circulated throughout the building. Solutions to this problem will be discussed later.

The final scenario when linking SBS to air conditioning systems is that contaminants passed throughout the air may have an allergenic effect on occupiers, i.e. once occupants are sensitized subsequent very small doses can cause illness.

To summarize this section, it is necessary to identify all the possible causes for SBS that can be linked to air conditioning. These include airborne pollutants, chemical pollutants, air-borne dusts and fibres and microbiological contaminants. These can be 'pulled' into the air conditioning units and distributed around the building. The number of air-borne pollutants in offices and similar environments is enormous and their sources can be numerous.

The pollutants released by the building occupants include CO_2, water vapour and microbiological organisms and matter. Cigarette smoke is also a pollutant, but the presence of this is due to the building occupiers. The many chemicals in cigarette smoke can be irritants. Many sources of pollution have been attributed to releases from the fabric and furniture within the building. These include formaldehyde, especially from urea formaldehyde insulation, organics and solvent vapours from adhesives used for furniture and sticking carpets, and dust and fibres from carpets, furniture and insulating materials. In a recent study two factors turned out to be important correlates of symptoms [12]: the 'fleece factor' (total area of carpet, curtains and fabric furnishing divided by the volume of the space), and the 'shelf factor' (length of open shelving or filing space divided by the volume of the space). These factors reflect the extent of possible causes of pollution such as organic, dust and microbiological elements. There is evidence that removing carpets may reduce symptoms.

There are, however, problems with an explanation of SBS in terms of pollutants. Several studies have examined variation in symptoms in relation to indoor air pollutants and the general conclusion has been that there is no difference in pollutant concentration between 'sick' and 'non-sick' buildings. Although the evidence on individual pollutants is inconclusive, SBS could result from the additive synergistic effect of many pollutants, each of which is individually subthreshold. If this is the case, then summary 'risk indicators' such as fleece and shelf factors may be more useful in some cases than measurements of specific pollutants. Photocopiers have been put forward as a possible cause of SBS, as pollutants such as ozone can collect in poorly ventilated areas.

Possible control of SBS symptoms thought to be caused by air conditioning systems

Recommendations have been made as to how SBS can be avoided within buildings that are mechanically ventilated [11].

1. A minimum fresh air flow of 8 litres per second per person is recommended. Where heavy smoking is permitted, an air flow rate of up to 32 litres per second is suggested.
2. Air velocities should be in the region of 0.10–0.15 m/s, rising to 0.25 m/s in summer.
3. Intakes for air-handling equipment should be located in such a way that they do not draw in air contaminated by traffic fumes or cooking smells.
4. Alterations to the layout of a building can affect the efficiency of naturally and mechanically ventilated buildings. Normally aspirated air should be designed so that the natural air movement is not adversely affected. For air conditioned buildings, the prime consideration is not putting up partitions that block either the intake or extract routes of the air supply.
5. There is no defined ideal temperature but the minimum recommended level is 16° C, with 19° C considered to provide a reasonable comfort level. Failure to control the temperature is unlikely, by itself, to cause SBS but it can influence other factors such as exposure to airborne pollutants.

6. Humidity should be maintained at 40–70%. In warm offices the relative humidity should be at the lower end of the range. Humidity, like temperature, is not on its own likely to cause SBS, but it can contribute to other symptoms. High humidity can encourage harmful bacteria, whereas low humidity contributes to a dusty atmosphere with the risk of dry eyes, nose, throat and skin.
7. Office equipment, such as photocopiers and printers, should ideally be located in a closed room with a separate extraction system.

HUMIDIFICATION EQUIPMENT

Various types of equipment are available for controlling humidity [13]. The risk of humidifier fever depends largely on the type of humidifier used and, in particular, whether water is stored and/or recirculated, and whether the humidifier is able to release water droplets. The main types of humidifier are:

- Cold water evaporators (Fig. 2.3): these rely on a wetted porous element over which the air flows. The element may be stationary and have water supplied to it, or it may be drawn over a drum or cylinder dipping into the reservoir. This type of system, although not associated with humidifier fever, is capable of releasing organisms.
- Hot water evaporators (Fig. 2.4): these contain water which is heated to release vapour. These are not associated with humidifier fever.
- Steam injection: this can be either from a central boiler or a self-contained unit. Steam injection is not associated with humidifier fever, and is probably least likely because of the sterilizing effect of steam.
- Compressed air atomisers (Fig. 2.5): these release a fine spray directly

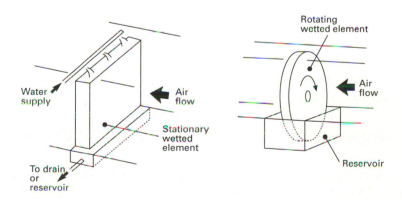

Fig. 2.3   Cold water evaporators.

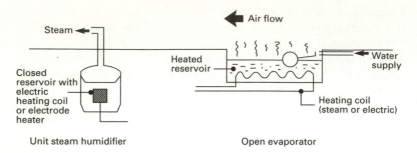

Fig. 2.4 Open hot water evaporator and unit steam humidifier.

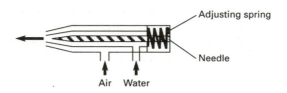

Fig. 2.5 Compressed air atomizer.

into the workroom from a series of nozzles fed from water and compressed air pipes. Water may be supplied from the mains or a tank. No cases of humidifier fever have ever been associated with this type of humidifier.

- Spinning disc atomizing humidifiers (Fig. 2.6): these are often mounted in the workroom area itself, and create a fine spray which in theory quickly evaporates. Water is often fed into a pan in the base of the humidifier which can become heavily contaminated, so there is a risk even when the humidifier is connected directly to the main water supply. This type of humidifier has been implicated in a number of cases of illness.
- Spray humidifiers and air washers (Fig. 2.7): these create a water spray, which is generally coarse. Baffle plates are fitted to prevent the release of water droplets into the ducts but these are of limited efficiency so finer droplets will inevitably escape. Cases of humidifier fever have also been associated with this type of equipment.

Precautions that need to be taken when installing humidification systems

The main object is to prevent the dispersion of heavily contaminated water droplets from humidifiers and air washers [14]. This can be achieved by the correct choice of humidifier system, good maintenance and regular cleaning. Where possible humidifiers that present the minimum

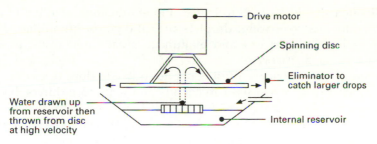

Fig. 2.6 Spinning disc.

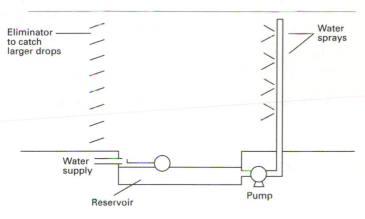

Fig. 2.7 Spray humidifier (or air washer).

risk should be chosen. Steam humidification presents the least risk as it does not create droplets and the steam may kill organic growths. However, running costs may be higher. It is not always safety that is the main prerequisite for the types of system to be installed; sometimes technical and cost requirements are the main priorities. This may lead to the installation of a system that is not as safe. Where it has been necessary to install such a system, other precautions assume greater importance.

- Water supply: the supply should be clean and free from contamination. Taking water from the mains greatly reduces the risks. If a tank is to be used for storage then it must be flushed, cleaned and filled with clean water regularly.
- Cleaning and disinfection: regular cleaning and disinfection of humidifiers and storage tanks is extremely important, especially if the system is of the spray or atomizing type. The frequency of cleaning will depend upon environmental conditions; more cleaning will be required in warm weather.
- Maintenance: a good standard of maintenance is required. Particular

attention should be paid to baffles and eliminators intended to mini-mize the release of water droplets and all types of humidifier should be well maintained to avoid malfunction that could result in water entering the ductwork.

- Water treatment chemicals: it is not advisable to dose reservoirs with these chemicals when used in conjunction with humidifiers, because of the potential danger to occupants. However, their use may be unavoidable in some circumstances where air-borne contamination by organic growth cannot be controlled using other methods.

Where cases of humidifier fever have already occurred it is often too late to take remedial measures with the existing humidifier, even with extensive cleaning and maintenance. If symptoms persist after these mea-sures have been taken there are three alternatives:

1. to move the sufferer away from the source;
2. to dispense with the humidification system (not practical in certain industries);
3. to replace the humidification system with one that has little risk; this may prove expensive but is the only practical solution.

DESIGN OF BUILDING SERVICES

The traditional route to the design of building services is shown below.

1. client advises architect of requirements for the building;
2. architect produces outline sketches for approval;
3. architect then passes on the outline sketches to the structural engineer who carries out structural design;
4. architect then passes on both sets of drawings to the services engineer.

The services engineer only becomes involved a long way into the design process. By this time the services engineer has to try to incorpor-ate all the client's mechanical and electrical requirements within the building shell. There are those who would argue that a better route would be to gather all the designers together at the earliest possible opportunity to discuss all aspects of the building. In this scenario the ser-vices engineer can design the air conditioning system to give the best results, and the building can be designed to accommodate this. The coor-dination of services on building sites has often been a problem through-out the construction phase, but early involvement of the services designer could go some way to alleviate this. If services are designed in isolation and tested as such, there is no evidence that when they are in contact with other components, or worse, moved to other positions, they will behave in the same way. There is a growing trend towards this early involvement of all the design team, especially in the design and build sector, and this should improve the overall performance of the services package.

Integration between all systems is of the utmost importance [15]. Elements such as heating, ventilation and lighting should be designed so that they do not impair one another's effective operation. For example, the mechanical and engineering layout should not interfere with the lighting layout. Designers should attempt to allow flexibility in the function of the building so that different uses can be easily mixed. For instance, if an office is occupied 24 hours a day, the system must be able to adapt to both night and day use.

CONSTRUCTION PRACTICE

As mentioned above, the installation of the air conditioning system must be carried out in such a way as to ensure that the system is not prone to breakdowns, is clean to start with and is maintainable. A paper produced by the HVCA in 1991 gives a guide to good practice with regards to the internal cleanliness of new ductwork installations [13]. The proposals range from a basic standard of cleanliness suitable for normal commercial/industrial buildings to higher standards where more stringent degrees of care and protection are required. Recommendations mentioned in the guide will only be applicable if the environment in which construction is being undertaken is also clean. Therefore some of the responsibility lies with the main contractor. In order to achieve the necessary levels of cleanliness the ductwork contractor requires first a dry, clean storage area, adjacent to the working area. During erection of the ductwork the working area itself should be clean and dry.

The following levels of cleanliness and protection have been identified.

Basic level: condition of ducts ex-works (as they leave the manufacturing plant)

Ductwork leaving the premises of the manufacturer will include some or all of the following:

1. internal and/or external self-adhesive labels for identification;
2. exposed mastic sealant;
3. light zinc oxide coating on the metal surface;
4. a light coating of oil on machine formed ductwork;
5. minor protrusions into the airway of rivets, screws, etc.;
6. internal insulation and associated fixings;
7. discoloration mark from plasma cutting process.

It must be noted that ductwork should not be wiped down or specially cleaned at this level. Ductwork delivered to site will have no special protection but must be protected against damage en route and during unloading. Before installation each duct is to be checked for debris before installation but not wiped. Openings in the duct do not need to be covered but the installation should be protected from damage at all times.

The designer must decide at design stage the number and size of accesses into the ductwork to ensure checking and maintenance can be carried out.

Intermediate level

In addition to the provisions of the basic level the following requirements should also be met:

- The area provided for storage shall be permanently clean, dry and dust free. This may require a boarded floor and water resistant covering.
- The working area must be clean and dry and protected from the elements, the internal surfaces of the ductwork must be wiped to remove excess dust immediately prior to installation and open ends on completed ductwork must be covered at the end of the work shift.

Advanced level

In addition to the provisions of the intermediate level the following requirements should also be met:

- All self-adhesive labels for identification should be fixed externally, and all ductwork during transport should be sealed either by blanking or capping and small components bagged.
- In addition to the provision of a clean, dry, dust free environment, all sealed ends must be examined and if damaged resealed with polythene.
- The working area shall be clean, dry and dust free. Protective coatings shall only be removed immediately before installation.

The three levels are summarized in Table 2.1.

Table 2.1 The three levels: Summary

Level	Factory seal	Protection during transit	Protection during site storage	Site clean	Cap off on site	Special clean
Basic	No	No	No	No	Risers	No
Intermediate	No	No	Yes	Yes	Yes	No
Advanced	Yes	Yes	Yes	Yes	Yes	No

SPECIALIST CLEANING

Where specific limits of cleanliness are required – and high levels of cleanliness should be specified to prevent SBS – ductwork shall be cleaned after installation by a specialist cleaning contractor. This will not normally form part of the ductwork installation and shall be specified separately. Accesses must be allowed for in the design to allow cleaning of the installation to be undertaken thoroughly.

The methods for cleaning the ductwork system are as follows:

1. vacuum
2. steam
3. compressed air
4. chemical
5. disinfection

The method chosen generally relates to the use of the building. Important considerations which also have to be looked at are that some sealants and cleaning materials can cause problems to the user. However, the COSHH regulations require employers to assess employees' health from using these materials.

Effects HVCA recommendations may have on the numbers of buildings that are prone to 'sickness'

It has already been proved, through research, that there is no concrete evidence to link SBS solely with air conditioning systems. However, improvements in the installation procedures for ductwork can only be of benefit. If all the guidelines are adhered to the installation should be spotless, undamaged and in perfect working order. There should be adequate access for maintenance and cleaning purposes now and in the future.

The choice of cleaning medium for the inside of ductwork must be carefully considered, as the use of chemicals or disinfectants may have an effect on occupants after cleaning is completed and the system is 'up and running'. These problems are not covered by the COSHH regulations.

THE SOURCES AND CONTRIBUTION OF NOISE TO SBS

Current evidence suggests that there is no one cause of SBS. However, it is suggested that it is caused by a number of factors which combine to produce the symptoms of SBS. Noise contributes to SBS, but in differing ways [3]. Low frequency noise can have a direct link with symptoms, whereas high frequency noise has been found to be inversely related to symptoms. Office equipment with low noise emissions can help dampen the general noise level in an office. Forms of intrusive noise that can be avoided by the careful design and rerouting of services include the sound of air passing through diffusers and ductwork and water in pipes and machinery such as lift motors or air conditioning plant.

The external fabric of the building needs to have noise insulation properties. This can be achieved by using multiple skin construction, which breaks the path of air-borne sound externally. By using raised floors and ceilings, structure-borne sound can be reduced. The type of furnishings to be used, and the type and layout of partitioning can be specified taking into consideration the need to cut down reverberation time. Both of these measures will cut down on excessive background noise.

If the problem in a bulding is low frequency noise from office equipment, this could be 'deadened' by introducing a level of background noise into the office.

PERSONAL ENVIRONMENTS SYSTEMS

In autumn 1988 Johnson Controls Incorporated introduced Personal Environments [16, 17], a system of individual environmental control for open plan office workstations. The system was installed in the offices of the Marketing Communications Department of the Controls Group of Johnson Controls Inc. in Milwaukee, Wisconsin. Prior to installation these offices displayed a typical office landscape, having been refurbished in 1984. The layout was designed by Johnson Controls in-house facilities department, in accordance with corporate standards.

The original layout was typical of many office settings today; the isometric drawing in Fig. 2.8 details the layout after installation of the Personal Environments System. There are 24 workstations in the

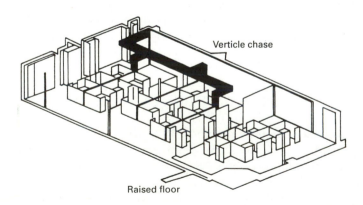

Fig. 2.8 Personal Environments demonstration office at Johnson Controls, Milwaukee, Wisconsin.

department each with a personal environment module (PEM) installed. Each workstation has access to facilities – electricity, telecommunications, data communications and conditioned air – through a vertical chase or access floor distribution method. The temperature of the common areas and aisles in the open plan office is not tightly controlled but allowed to float. These areas are conditioned by the 'spill over' from the PEMs in the workstations, using a 'loose fit' concept. The principle behind this system is that individuals are able to tune or adjust the systems to meet their own needs. Heating is provided by a conventional hot water radiant system. The general air flow patterns of the office before and after installation of the PEMs are shown in Figs 2.9 and 2.10.

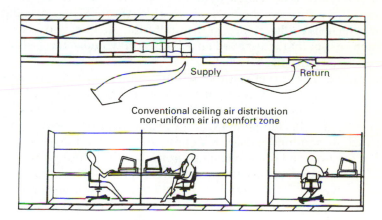

Fig. 2.9   Air flow pattern before the installation of the PEM.

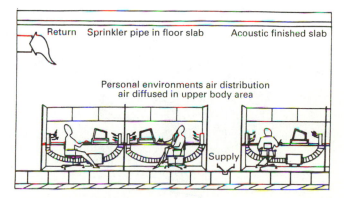

Fig. 2.10   Air flow pattern after PEM installation.

PEM

The basis for the Personal Environmental System is the PEM. The PEM is one of the five key components which makes up the workstation system; the others include a desktop control panel for the user controls, two desktop diffusers for air distribution and a floor-standing radiant heat panel. The PEM is designed to mount beneath the work surface of open plan furniture systems. The PEM has two air inlets, one for conditioned air (supply) and another for room air (return). The output (exhaust) air temperature is a blend of supply and return air. The PEM is a cooling only system.

Air flow is provided by fans installed in the PEM which are chosen for their low noise emissions, smooth running and reliable operation. The speed of the fans is adjustable by the user to vary the flow of the supply air from the diffusers. The air is discharged through two desktop diffusers which have the ability to distribute the air in a wide variety of directions as set by the user. All air – both supply and return – is drawn through an electrostatic air filter.

The radiant heat panel is a free standing unit which allows the occupant to position it at any convenient below desk location, based on preference or work habits to provide heat for the lower body; the user has the ability to control the temperature of the heat panel from the desktop control panel. Background noise can be provided through the use of a white noise generator. The noise emanates from the PEM through the air diffusers and can be set to whatever intensity provides the occupant with a sense of privacy. A light control also located on the desktop control panel allows for adjustment of the lighting to the intensity required for the particular work being done. An infrared sensor in the control panel automatically turns the PEM off and on when the station is unoccupied or occupied. Figure 2.11 shows the main features of the PEM.

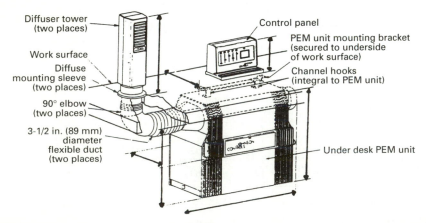

Fig. 2.11   The main features of the PEM.

Figures 2.12, 2.13 and 2.14 show the section and plan views for the Personal Environments access floor application, and also a sectional view of the raised floor connection.

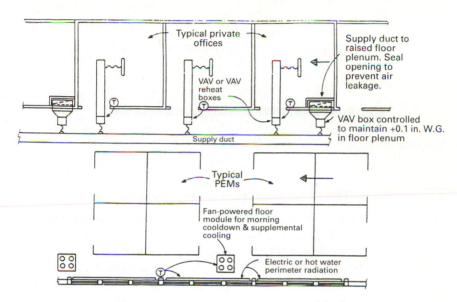

Typical private offices

VAV or VAV reheat boxes

Supply duct to raised floor plenum. Seal opening to prevent air leakage.

Supply duct

VAV box controlled to maintain +0.1 in. W.G. in floor plenum

Typical PEMs

Fan-powered floor module for morning cooldown & supplemental cooling

Electric or hot water perimeter radiation

Fig. 2.12   Personal Environments access floor application – plan view.

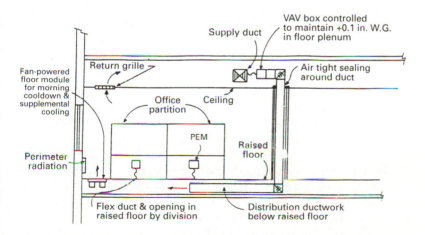

VAV box controlled to maintain +0.1 in. W.G. in floor plenum

Supply duct

Return grille

Air tight sealing around duct

Fan-powered floor module for morning cooldown & supplemental cooling

Office partition   Ceiling

PEM

Raised floor

Perimeter radiation

Flex duct & opening in raised floor by division

Distribution ductwork below raised floor

Fig. 2.13   Personal Environments access floor application – elevation/section view.

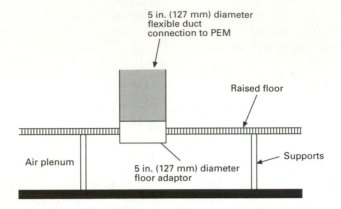

Fig. 2.14   Sectional view of raised floor connection.

Results obtained by using the Personal Environments System

Once installed the operation of the facility was examined from two different aspects. One of these was the use of the PEMs themselves. A digit controller included in each PEM formed the interface between the occupant and the PEM equipment. One of the features was a built in data communication capability. This capability helped create a network of PEMs. This system was used to gather data at five-minute intervals from each PEM. The data included occupancy, temperature and air flow. The results proved very interesting; the temperature range was recorded as 64.4°F to 78.8°F. This shows how extreme personal preferences can be.

From the results obtained from these tests we may make some assumptions. If we assume that one of the contributory factors of SBS is the fact that everyone's personal temperature, lighting, etc. requirements are different, then it is impossible to design a central HVAC system to suit everybody, although by using average values we may satisfy a majority. This may be a contributory factor to the fact that not everyone within an environment will suffer from SBS, and that it affects different people to different degrees. This poses two questions: 'Are those who are unaffected by SBS, those for whom the central HVAC provide suitable conditions?', and 'Are those worse affected those who would choose extreme conditions for their personal environments?' If this is the case then the use of PEMs may be one way to eradicate SBS from the workplace.

Other benefits were perceived by the users of these systems. Visual quality improved and fewer people were distracted by others talking. Overall the occupants felt the office was healthier and had a positive effect on their work. The occupants tended to stay for longer periods at their desks, increasing productivity. System performance was monitored and documented by a team of experts. A functional analysis expert through a series of interviews obtained the comfort perceptions of the occupants of this installation. He asked the same questions in June 1988 before the PEMs were installed, and then again in October 1988 after installation. Some of the results are shown graphically in Figs 2.15, 2.16 and 2.17. This is by no means a full set of results, but the results for a vast range of environmental factors including noise, smoking and air quality were highly favourable to the use of the PEM.

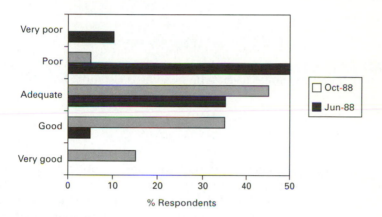

Fig. 2.15 Answers to the question 'Generally at this time of year, how would you rate the temperature around your work space?'

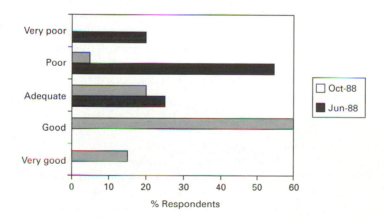

Fig. 2.16 Answers to the question 'Generally at this time of year, how would you rate the air freshness around your work space?'

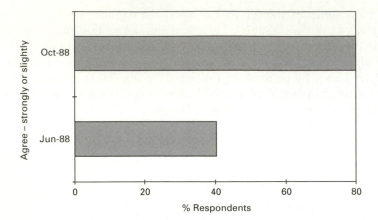

Fig. 2.17   Responses to the request 'Please indicate whether you agree or slightly agree with the following statement. "This building is healthy to be in."'

INDUSTRIAL CLEANING OF THE WORKPLACE

Cleaning and maintenance are two of the most important factors influencing SBS [11]. Regular cleaning, undertaken after office hours when dust disturbance has time to settle before the building is occupied, is recognized as a major boost to office morale. Likewise, regular maintenance of items as diverse as the air conditioning, building fabric, lighting and lifts creates a sense of well-being among employees; it shows that the company's management appreciates them and is concerned to create a pleasant working environment.

   An untidy workplace looks unprofessional and can create a safety hazard. It has been estimated that if workplaces were kept as clean as domestic dwellings, up to half the accidents that occur could be eliminated. This could also account for some of the causes of SBS. Janitorial services for office buildings are often way down the list of priorities for many companies. A survey carried out by Dyna-Rod concluded that the majority of building owners only carried out crisis maintenance; they waited for the system to fail before carrying out any repairs. Despite its rather mundane and less than glamorous image, industrial cleaning is a specialized industry, requiring trained staff, better equipment, detergents, etc. One company, Clean World Environmental Ltd, launched a new industrial cleaning service to comply with the COSHH regulations in 1989 [18]. The service involved specializing in the cleaning of tanks and plant equipment which had been exposed or used for hazardous substances and which could contaminate the environment, especially the work environment.

   There are a multitude of industrial cleaning organizations which can carry out, or advise on, cleaning procedures in buildings of any size and complexity. By using these organizations there may be a reduction in the symptoms of SBS if pollutants are said to be the main cause, although this has been refuted.

Even if there is no evidence to conclusively prove that dust and air-borne pollutants are the sole, or even a contributory, cause of SBS, we can assume that housing employees in a clean environment will almost certainly boost morale. For this reason it is almost certainly a good idea for the employer to obtain the advice of a specialist cleaning firm.

CONCLUSION

Having looked at the evidence for each suggested cause of SBS, it begins to look like an effect without a cause. No single factor can be identified as the major cause of the problem, and evidence has been put forward to prove that each of the causes thought to be responsible for SBS is not actually responsible. The British Research Establishment has assembled a multidisciplinary team of researchers who will push forward the search for a cause and then a cure. As part of the thinking behind this research they will make two important departures from most research that has gone before.

First, there have been no large-scale attempts to carry out a systematic modification of environmental conditions. Most of the previous investigations have focused on comparisons between buildings, which will always complicate procedures as the many factors involved in a building make comparison difficult. These studies have provided important clues as to the cause, which could now be tested by a double blind examination of changes in symptoms from varying environmental parameters.

Secondly, no single study has examined all the likely causes as part of one study. Such a study would make it possible to identify interactions among factors and to take account of different causes in different buildings. If there has been little by way of assessment of remedial measures, still less has there been clear demonstration of the cure of a sick building. Claims are made about cures but it is difficult to substantiate these in the absence of controlled follow up investigations.

SBS is not an isolated phenomenon. There are many complaints of SBS in different countries. A World Health Organization working group report stated that although the incidence of SBS varied from country to country, up to 30% of new or remodelled buildings have an unusually high rate of complaints. This estimate is to some extent arbitrary, and could be set even higher by taking a different definition of a high rate.

The consequences of SBS are reduced work performance and increased absenteeism, and even building closure, as in the case of the Inland Revenue building in Liverpool. In addition, because SBS may become associated with energy conservation measures it could, probably unjustly, militate against their wider implementation. One result is the value of increasing ventilation (energy costs) to benefit from improved staff productivity (decreased staff costs). There are other problems which occur morally if we increase ventilation; the results are increased energy costs, loss of non-renewable energy sources and increased carbon dioxide emissions which contributes to the problem of global warming. Full consideration should therefore be given to natural ventilation alternatives to air conditioning.

Sick Building Syndrome is neither life threatening or disabling; but the symptoms are seen as important by those who suffer from them. With the predicted movement from 'blue collar' to 'white collar' employment in the future, and the majority of those affected being 'white collar' workers, finding a cure for SBS is of paramount importance.

REFERENCES

1. Sykes, J.M. (1988) Sick building syndrome: a review, paper from the Health and Safety Executive.
2. World Health Organization Regional Office for Europe (1982) *Indoor Air Pollutants: Exposure and Health Effects: Report on a WHO Meeting*, Norlinger, 8–11 June.
3. *CIBSE Guide*, Chartered Institution of Building Services Engineers, Delta House, 22 Balham High Road, London SW12 9BS.
4. Gilperin, A. (1973) Humidification and upper respiratory infection incidence, *Heating/Piping/Air Conditioning*, March.
5. Green, G.H. (1984) *Studies of the Effect of Air Humidity on Respiratory Diseases*.
6. Green, G.H. (1984) The health implications of the level of indoor air humidity, *Indoor Air*, **1**, Swedish Council for Building Research.
7. International Standard ISO 7730-1984. Moderate Thermal Environments: Determination of the PMV and PPD Indices and Specifications of the Conditions of Thermal Comfort.
8. Wojtas, O. (1989) *Glasgow Herald*, 6 September.
9. *The Economist* (1989) Architect, heal thyself, 13 May.
10. BSRIA (1989) *Environmental Comparison of Air Conditioned and Non-air Conditioned Buildings*.
11. Rideout, G. (1995) Itching to go home, *Building Services*, 23 June.
12. Raw, G.J. (1989) *Sick Building Syndrome*, Research Project EP244.
13. HVCA DW/TM2 (1989) *Guide to Good Practice: Internal Cleanliness of New Ductwork Installations*.
14. The Health and Safety Executive (undated) *Report on Humidifiers*.
15. McNeil, J. (1995) 'Place ache', *Building*, 12 May.
16. Johnson Controls Incorporated (1989) *Personal Environments: 'The Office of the Future'*, Project Report.
17. Johnson Controls Incorporated (1990) *Personal Environments Application Manual*.
18. Greek, D. (1990) Sweeping changes at work, *Professional Engineering*, January.

CHAPTER 3

Light and lighting

Hana Drahonovska

Light is a part of our natural environment, like air and water, or a component of our artificial environment in buildings. Lighting is light used for the comfort and activity of people and, like heating and ventilation, can be controlled by technical means.

Lighting is related to both general satisfaction in the indoor environment and the comfort of visual performance. Eye work under inappropriate lighting can be a very obvious cause of sick building syndrome (SBS), producing eye discomfort, eye strain and fatigue.

Although no direct damage to the eyes or visual pathway has been confirmed by researchers or occupational physicians, the biological effect of light not only on visual task performance but also in controlling most of the physiological and psychological functioning of human organisms is evident.

Light and lighting as perceptual factors are strongly connected with the development of cognitive and emotional functions. We receive nearly 90% of incoming information by the visual pathway – for orientation, learning and working. The last decade has brought the widespread use of computers and other VDUs that change the demands of visual work. The eyes are the most important working tool across a wide spectrum of jobs and professions, and they have to function in an appropriate lighting environment.

Recent research has focused on the relational/perceptual approach rather than the classical photometric approach because the ultimate effect of light and lighting on human health and well-being depends on individual perception and satisfaction.

However, higher illuminance, higher visual performance and less eye strain are related to a reasonable eye response to visual work; it is necessary to know how lighting causes eye strain.

Stress is commonly taken as a negative term implying acute or chronic impairment of health. Stress as a term in ergonomics is defined simply as an environmental factor with changing characteristics. The reaction of the human body to any stress is expressed by the term strain.

There are two main types of reaction to stress:

1. The greater the stress, the greater the body's biological response. To decrease strain is necessary to minimize the stress factor.
2. Strain tracks stress, both up and down. To decrease strain it is necessary to optimize the stress factor.

Lighting should be considered as a stressor of the second type. Daylight is at least a hundred times more bright (100 000 lx in summer, 20 000 in winter) than typical artificial illumination (less than 1000 lx). However, complaints about 'too bright' artificial lighting at levels of 500 lx are as frequent as 'too dim' daylight at levels of 5000 lx. For this reason artificial lighting norms ensure the lowest limits of artificial lighting but not the highest. Although it is generally agreed that more lighting is preferable, this is not true for all types of lighting, people and tasks.

Symptoms of SBS related to lighting can be caused by both environmental and human factors – as it is shown on the Table 3.1.

ENVIRONMENTAL FACTORS

Light and lighting

Light is electromagnetic radiation with a wavelength spectrum and is divided into three regions – ultraviolet, visible and infrared light – each having a different effect on humans. Ultraviolet light (UV) is radiated in the region 250–380 nm and has three different spectra labelled UVA, UVB and UVC affecting the cells of the human body at different levels. The

Table 3.1 Symptoms and causes of SBS

Symptom	Environmental cause	Human factor
Eye discomfort	Lighting Air pollutants High temperature Low humidity Allergens ETS	Lack of sleep Eye lens wearer Smoking Eye disease Hypersensitivity Allergy Photosensitivity
Asthenopia	Lighting	Eye defects Long visual work
Eye strain	Lighting Noise Other factors	Psychological profile Hormonal imbalance Depression

main affected organ is skin. Health risk from damage by UV radiation has been discussed in many professional and public fora. The probability of health damage caused by UV light indoors is very low, for the following reasons:

Window glass absorbs at least 70% of sun UV radiation. Lighting covers also filter out UV lighting from artificial sources. In fact some lighting device manufacturers use special glass to absorb nearly 100% UV radiation. In discussing the effect of UV radiation indoors it is more appropriate to consider its benefits. The positive effects of UV light include a lowering heart rate, increasing the metabolic rate and general activity, a quicker reaction to sound, light and other stimuli, less fatigue of the visual receptors, better resistance to some respiratory tract infections and synthesis of vitamin D, promoting the metabolism of calcium and phosphorus in the human body. Briefly, UV light has in general positive physiological and psychological effects on humans.

Visible light is radiated at 380–760 nm. The spectrum includes different colours related to the wavelength (Table 3.2).

Visible light enters the human body by both eye and skin. Light entering through the skin has a small effect in comparison to ultraviolet or infrared light and it can be disregarded by the majority of the population except for a few photosensitive people who react by rash, itching, redness and other skin allergic complaints. Photosensitivity of the eyes such as photophobia and intolerance of normal brightness can impair vision.

Visible light entering through the eye is changed through biochemical processes into a neural sensation using the visual pathway. Neurofibres leading to the visual cortex to mediate vision are called the optical portion of visual pathway, while neurofibres finishing in the hypothalamus (energetic portion) as the target organ, induce and control biological function in animal and human organisms. Both parts of the visual pathway were described by Hollwich [1] to explain the effect of invisible light entering through the eye.

Infrared light (IR) is radiation over 760 nm and has a predominantly thermal effect. It can penetrate deeply into skin structures and muscles,

Table 3.2 Spectral composition and colour of light

Colour	Wavelength (nm)
Violet	380–455
Blue	456–492
Green	493–575
Yellow	576–585
Orange	586–647
Red	648–676

resulting in increased body temperature and blood circulation, thus influencing physical and mental performance. The radiation of human skin by IR can be considered as a regulator of general activity level. Effects on the eye can be related to high temperature, causing destruction of proteins and possible damage of eye tissues. Extreme thermal destruction of the retina could lead to blindness.

Daylight and lighting

Daylight comes from either direct or reflected sunlight and varies in both quantity and quality, changing with time of day, season and weather. Human eyes are phylogenetically adapted in both spectral distribution and amount of daylight; naturally occurring variances stimulate visual processes. The uniformity of artificial lighting causes 'boredom' of the visual structures and earlier onset of fatigue. For all these reasons daylight is the best lighting for visual work and well-being.

The importance of daylight for humans can also be evaluated from the following factors:

• biological and physiological (influence of biological functions of organisms);
• psychological (mood, activity, emotion).

Daylight is the most important regulator of chronobiological – circadian rhythms. Regular secretion of melatonin, the sleeping hormone released from the *corpus pineale* (pineal body) in the hypophysis, is controlled by the diurnal pattern of light and dark. Ashoff describes melatonin as *Zeitgeber*, meaning 'time giver' or 'time keeper'. Light-controlling melatonin levels are the underlying principle of the biological functions of organisms; however, the human biological clock is modified by social habit and style.

From the neurological viewpoint, light is a complex of signals that influence neural activity. These may trigger the central nervous system (CNS) to respond, since the human cognitive and emotional functions are strongly influenced by light.

Artificial lighting

From review articles one might conclude that artificial lighting is a poor substitute for daylight, although social life and habits necessarily need artificial lighting, providing stimulation for eye performance. However, artificial lighting should be used when insufficient daylighting is caused by season, night or weather, but not used permanently in windowless buildings. If a mixture of lighting is provided, the sensation of daylight should dominate.

There are several basic measurable parameters of artificial lighting:

Illuminance is the incident luminous flux per area unit measured in lux (lx). When illuminance is measured horizontally, it is called horizontal illuminance; the illumination of vertically oriented objects (walls, shelves) is called vertical illuminance. The measurements are taken using illuminance meters and readings are made at grid points at 85 cm above floor level. The number of grid points and their spacing is defined in national regulations and guidelines and varies with lighting design, type of work and room size. Special attention is recommended to individual workplaces, but the illuminance of the overall interior must always be measured.

Since the norms or guideline values of nominal illumination are not related directly to the satisfaction of occupants, illuminance alone should not be used to evaluate the effect of lighting but only to assess the technical criteria of lighting.

Luminance is the term for expressing the impression of brightness of a given object and is measured in candela/m^2 (cd/m^2). Luminance is a property of an object which differs and depends on position of observer; for example, the luminance of glossy objects is apparently different from that of matt objects. Glare, contrast and shadowness are assumed to be functions of luminance perception.

Measurement should be made (using photometers) at all or selected representative workplaces in the line of sight of the usual working position. Variations of luminance at the workplace and its surrounding are important for comfortable eye performance. The following luminance values are measured :

• of visual objects (e.g. screen, keyboard, sheet of paper);
• of immediate surroundings (e.g. desk in the middle and at the base);
• of more distant surrounding (e.g. wall, ceiling, floor, windows).

Glare is disturbance due to an excessively high degree of luminance or excessive variation of luminance in the visual field. Direct glare refers to glare resulting from objects with high luminance and sources of light: sun, lamps, visible sky. Reflected glare, often in offices, is caused by reflectance of bright objects and surfaces – windows, glossy desk, ceiling and floor – and can be removed by use of matt surfaces on visual objects. However, window glass, screens or glass covers over lighting sources will always be possible sources of glare. Arrangement of the workplace to avoid glare can be achieved only in small limited areas.

Glare (both direct and reflected) is perceived in three ways: (1) discomfort glare related to disturbance of concentration without visual impairment; (2) disabled glare which is followed by instant inability to undertake visual work; and (3) blinding glare causing lasting disturbance in vision even after the glare is removed from the visual field. Generally, office environments display small but very high radiant sources of glare (lighting, high glossy objects) causing discomfort and disabled glare.

Blinding glare is mostly caused outdoors from direct or reflected sunshine (e.g. large snow or sand areas).

Actual glare perception and its impact on vision depends generally on the following characteristics of the source:

- luminance
- size
- location on visual field
- contrast of background

The stronger the characteristics, the more impaired is the perception. Tiredness, nervousness and lack of well-being increase sensitivity to glare.

Veiling glare appears under certain conditions: light is reflected from dust particles, water, aerosols, dirty windows, etc. Veiling glare is rare in offices, but common in industrial plants with excessive dust or vapour pollution.

Contrast is the characteristic of luminance enabling visual objects or details to be distinguished from their surroundings. Under extremely poor lighting condition the contrast decreases to zero. The contrast rendering factor describes the relationship between contrast and referential contrast. Ideally, the same contrast rendering factor is maximized at all points in the visual field. Unfortunately this criterion can be difficult to achieve; contrast is often neglected by lighting designers.

Shadowness is the value expressing the modelling of visual spatial objects. Shadowness is a scalar/vector ratio depending on the direction of illuminance expressed in the range 0.5–3.0. For example, human faces have the most pleasant appearance under lighting with values 1.2.–1.8. Higher values are too harsh for human features; lower values cause shadow-free faces.

Shadowness influences the feeling of pleasantness or unpleasantness and general satisfaction with lighting. Shadows in the visual field are necessary to arouse the peripheral parts of the retina which relate to three-dimensional (spatial) perception. Uniformity of lighting stimulates fewer elements of the retina (less neural transmission) and decreases visual performance.

The colour of lighting is defined by the spectral radiation. If light is radiated at all wavelengths (polychromatic) it is perceived as neutral. If light radiates mostly in one wavelength (monochromatic), light is perceived to be unicoloured (blue, red, yellow, etc.).

The colour of light is expressed as chromatic temperature in kelvin (K). At a high chromatic temperature (e.g. 8000–10 000 K) light appears blue or white, while lower chromatic temperatures result in orange or red coloured light (800–900 K). White colour has a chromatic temperature of 3000–5000 K. The preferred supplement to daylight is warm white (near 3300 K) artificial lighting. The chromatic temperature of an incandescent lamp is 2500–2900 K, of a halogen lamp 3000–3300 K, and of fluorescent tubes 3000–6500 K.

Chromatic temperature influences the colour rendering index (Ra), as measured by the effect of illumination on colour appearance of an object by comparison with a colour appearance under a referential illuminant. Lighting for office spaces, for example, should not have values less than 80.

The influence of chromatic temperature (colour) of lighting on well-being also depends on the intensity of illumination as shown on Table 3.3.

Type of lighting

Each type of lighting has advantages and disadvantages. The choice of the best possible lighting for an environment depends on several characteristics of rooms – including quality and colour of surfaces, type of work, number of workplaces, size and height of room and orientation of windows.

Lighting is basically divided into two categories: general lighting to provide enough light in the room and task lighting providing light for the workplace. The use of task lighting is not common in office lighting, since the general lighting installed according to technical norms is unfortunately thought to be sufficient for human psychological and physiological demands. The most common types of lighting are:

- direct lighting (downlighting);
- semi-direct lighting (half downlighting);
- semi-indirect lighting (half downlighting);
- general diffusing (multidirectional);
- indirect lighting (uplighting);
- indirect lighting with task lighting (uplighting with local lighting, two-component lighting).

Figure 3.1 shows the types of lighting.

Table 3.3 Illuminance, chromatic temperature and perception

Illuminance (lx)	*Chromatic temperature (K)*		
	Warm (< 3500)	Neutral (3500–5500)	Cold (> 5500)
< 500	Comforting	Neutral	Cold
1000–2000	Stimulating	Comforting	Neutral
> 3000	Unpleasant	Stimulating	Comforting

		Direct	Semi-direct	General diffusing	Semi-indirect	Indirect
Percentage of light Approx.	Up	0-10	10-40	40-60	60-90	90-100
	Down	90-100	60-90	40-60	10-40	0-10

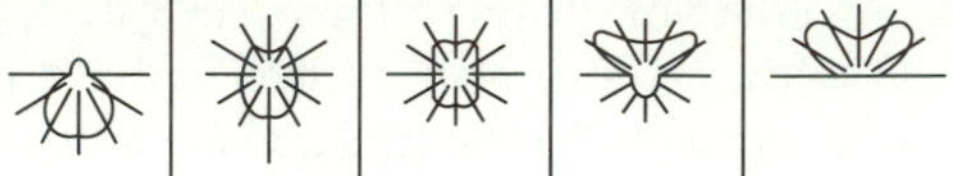

Fig. 3.1   Types of lighting.

Direct lighting: downlighting

Direct lighting uses ceiling mounted lamps with polished reflector grids with different angle-distribution characteristics. If a narrow angle is used, the lighting is called VDU lighting. This system is recommended for rooms with an internal height over 3 m. All working surfaces should be matt and workplaces must be arranged in relation to location of lamps. The illuminance flux is directed downwards only.

The advantages of direct lighting are:

• avoidance of direct glare from lamps;
• good illumination of workplaces if correctly planned;
• good illumination of the indoor space of a room.

The disadvantages are:

• lack of flexibility of position of desks or other workplaces;
• reflected glare if glossy surfaces are used in a room;
• lack of personal control of lighting according to individual task or to level of daylight.

The application of this type of lighting is recommended for workplaces with VDUs; but the satisfaction of VDU workers does not correlate with technical expectations.

Semi-direct lighting

This uses reflector grid luminaries with a narrow angle in the lower hemisphere and a wide angle in the upper. More than 60% of the light produced is directed downwards. The main advantages are good avoidance of glare, both direct and reflected. The disadvantages are:

• little flexibility of workplace position;
• lack of personal control over lighting;
• a feeling of reduced room height.

Multidirectional, general diffusing lighting

Multidirectional lighting uses round lamps mounted at different distances from the ceiling. Ideally, 50% of light is directed downwards. The main advantage is the good overall illuminance. The disadvantages are:

- direct and reflected glare;
- lack of personal control over lighting at the workplace.

Semi-indirect lighting

Semi-indirect lighting uses reflector grid luminaries with a narrow angle in the upper hemisphere and a wide angle in the lower. Less than 40% of light is directed downwards. The advantages are:

- avoidance of direct glare;
- strong reduction of glare disturbance;
- flexibility if mounted in each workplace.

The disadvantages are:

- no flexibility if mounted on the ceiling;
- little individual control by occupants.

Indirect lighting (uplighting)

Luminaries radiate light directed towards the ceiling or towards the walls. This system is recommended if glossy objects are used or shadows must be limited. The main advantages are:

- no direct glare, few shadows;
- good personal control if placed at each workplace;
- enhancement of the appearance of the human face.

The disadvantages are:

- high wattage followed by increased room temperature;
- lack of personal control if mounted on the ceiling.

Two-component lighting

The first component, indirect lighting, is used as general lighting, while the second component, direct local lighting, is used as task lighting. This is considered as optimal for both perception and visual performance and for the satisfaction of occupants. The advantages are:

- flexibility and adaptation to different tasks;
- adjustability to the individual needs of occupants;
- avoidance of direct and reflected glare;

- balanced vertical and horizontal luminance and illumination;
- good individual control over lighting.

The disadvantage is the requirement for a solid and bright ceiling and adequate room height.

Colours and decor of indoor environment

Colour is the attribute of spectral distribution of lighting. Colour perception is an important part of sensation, orientation and discrimination by visual processes and is generally based on three properties of colour: lightness, saturation and hue. The colour of decor, walls, flooring and other surfaces is related strongly to the well-being, emotions and satisfaction of the occupants. A brief summary of psychophysiological effects includes the effects of:

- warm colours (red, orange, yellow). These support activity, energy, performance and food consumption, but can also lead to fatigue. The application of warm colour is recommended for cold, north-oriented rooms;
- cold colours (blue, green, violet). These support relaxation including eye relaxation, peacefulness, mental concentration and creative thinking. Application is recommended in warm south or south-west oriented rooms;
- light colours, which create a feeling of space and brightness;
- dark colours, which create a feeling of closed space, anxiety and even depression.

Perception of colours depends on personal preferences, culture, habits, geographical location, nationality, personal physiological and psychological status, as well as type of work, indoor temperature, weather, matt or gloss surfaces and quantity and quality of lighting. It is strongly recommended that the main decor colour and furniture be discussed with occupants before refurbishment and painting.

The reflection of surfaces followed by reflected glare depends not only on matt–glossy characteristics of materials but also on colour lightness and hue. Darker surfaces absorb a greater proportion of light radiation, while lighter surfaces reflect a greater proportion. The most reflecting is white, the most absorbing is black. Other colours reflect light in decreasing sequence: yellow, green, pink, grey, blue, red, brown.

Other indoor environmental factors

Temperature

Inappropriate temperature can cause stress, including eye strain. Lighting is perceived to be brighter when indoor temperatures are above the recommended values. High temperatures also increase the emission of chemical pollutants from furniture, flooring and other possible sources, causing eye irritation that can be misinterpreted as caused by visual work in a poor lighting environment.

Relative humidity

Low relative humidity (dry air) can cause eye irritations.

VOCs

Formaldehyde and other volatile organic compounds emitted in the indoor environment irritate eyes, nose and throat, even at low concentrations.

Particles, fibres

Electromagnetic radiation around computer screens attracts dust particles and fibres. The level of the pollutants can be higher locally, causing eye irritation and other complaints that could incorrectly be ascribed to impairment caused by use of computers.

Allergens and air-borne micro-organisms

These can cause eye irritation under conditions of individual hypersensitivity.

Psychosocial environment

A lack of interest by managers, poor personal work skills, an inadequate pace of work, the quality of interpersonal relationships, job insecurity and lack of control of the indoor environment significantly affect a number of reported complaints.

HUMAN FACTORS

Visual system

Anatomy

The eye consists of three layers. The outer layer includes the sclera, the transparent cornea and the pupil; the middle layer is composed of the choroid, ciliary body, iris, dilator and sphincter pupillae and the lens in capsule with suspensory ligaments. The inner layer includes the retina and consists of receptors (rods and cones) and two important spots: the *macula lutea* (the yellow spot; its centre, *fovea centralis*, is the site of acute sight vision), and the optic disc (white or blind spot without any visual receptors) the origin of the optic nerve leading to the hypothalamus and the visual cortex (see Fig. 3.2).

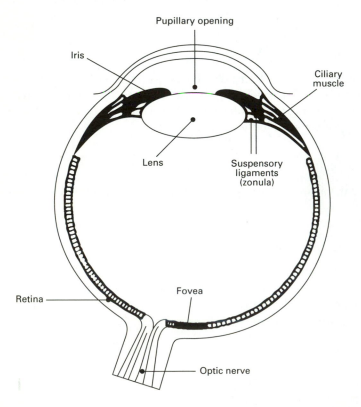

Fig. 3.2   The human eye.

Physiology

The visual system reacts to luminance and colour differences in the field of view. Light enters the eye through the pupil and the size of pupil is controlled by two muscles: the sphincter pupillae causes the iris to close, reducing the pupil diameter; and the dilator pupillae causes the iris to open, making the pupil larger. Under stable lighting conditions (especially luminance) the pupil is stabilized, although this state means constant small alteration in the range of 10% of the average diameter. These continual physiological fluctuations require strength of both muscles and can cause eye strain.

The pupil diameter depends on the light. Bright light improves visual acuity because the pupil becomes smaller and light can be better focused on to the central part of the retina. Dim light causes a larger pupil size, increasing the accommodation strength needed to focus on a visual object. The effects of the eye's refractive errors can be more obvious in such a light.

The pupil size is a function of not only luminance and distance of the field of view but also other circumstances. Mydriasis (large pupil size) corresponds with emotions, affection and anxiety, while myosis (smaller pupil size) corresponds with relaxation, concentration and thinking. Women and those with pale skin often have physiological mydriasis.

The variation in distance from visual work causes changes in pupil size more frequently than difference in luminance.

Accommodation is a crucial process for vision to define distances. This process uses contraction and relaxation of the ciliary muscle, which controls the refractive power of the lens. Maintaining the focus on a near object needs an increasing refractive power of the lens due to the continuous contracting force. Accommodation efficiency and range decreases with age. The nearest point (the smallest distance at which the object is seen sharply) is 10 cm from the eyes at age 15 but 13 cm at age 30 and 50 cm for 50 year-olds.

Adaptation

The retinal receptors (cones and rods) generate electrical signals to the brain via chemical processes and they are translated back to visual perception. The level of adaptation (or luminance sensitivity level) of the retina is the key to efficiency and vision comfort.

There are two types of adaptation based on chemical transformation of the retinal pigment rhodopsin, followed by pupillary size response. Dark adaptation increases luminance sensitivity and has a duration of 30 minutes, while light adaptation is a decreasing luminance sensitivity response and takes 5 minutes.

Control mechanisms for instant adaptation to small changes in luminance are based on electrical signals from retina receptors. This complex process is locally based, where the more active receptor inhibits the adapting activity of nearby receptors. The eye adapted to the actual luminance in the visual field is comfortable only if the luminance is relatively constant in working space and time.

Visual acuity resulting from pupillary response and lens accommodation is the ability to see two points as being separate. The examination of this type of acuity, termed 'minimum separable acuity' is the most frequently preferred method in ophthalmology.

The ability to detect a discontinuity in a line, called Vernier acuity, is less commonly used. Examination of this type of visual acuity requires different tools and its sensitivity is about 10 times less than minimum separable acuity.

It is generally accepted that visual acuity is better with increasing illuminance – but this is only partially true. The speed of reading increases by about 20% if illumination is increased from 20 lx to 100 lx. A further increase in illuminance, however, has less effect on the speed of reading and illuminance over 500 lx has little effect. Illumination over 1000 lx, often recommended as the optimum lighting level for office work, has negligible influence on speed and accuracy in recognizing details of written words; in fact it can cause early symptoms of eye astenophia, resulting from fatigue.

Visual acuity depends on either the quality of optical eye media or proper correction of errors of refraction. Correct glasses require knowledge of the work distance: the main visual task distance is different for conventional pen and paper work (30–35 cm) and for computer work (50–60 cm).

Colour vision

The three types of receptors in the retina are differentially sensitive to red, green and blue light; hence perception of colours depends on the complex variations of signals from the receptors to the brain.

Anomalies of colour vision are either inherited or acquired and can be caused by impairment of any part of the eye or visual pathway. Diabetes mellitus, chronic abuse of alcohol, prolonged exposure to organic solvent or lead are some of the more common causes of acquired disturbance of colour vision.

People with colour deficiency tend to make more errors during a visual task or need more time to perform it.

Colour discrimination also decreases under low levels of illuminance.

The visual system is not equally sensitive to all wavelengths of visible light. The most sensitive wavelength is 545 nm (yellow-green light) with a smaller secondary peak at 610 nm (orange-red light). Different wavelengths focus on the retina at different distances; for example, blue light is refracted more than red light. These chromatic aberrations can cause eye discomfort due to the extra strength of the ciliary muscles required to accommodate red lighting or a predominantly red environment. On the other hand, blue light and blue colour can be an advantage for accommodation, except that the lens becomes more yellow with age, becoming less transparent to blue light. The spectral sensitivity of the elderly is shifted toward longer wavelengths (red lighting). Chromatic aberration is a more severe problem with very low illuminance (below 100 lx).

Visual performance is defined as the speed and accuracy of visual work and is usually tested as the number of mistakes made during a fixed time period. Results are expressed as amount of information transmitted to the brain's visual centre (in bytes).

In the past it was believed that visual performance was improved under higher level of illumination; but it has recently been shown that this is only partially true for central vision. Central vision is vision caused by arousal of the visual elements in the central – foveal – part of the retina, which is the most important for distinguishing details. Peripheral retinal vision is based on shadows and brightness and the three-dimensional relation of visual objects in the whole field of vision.

Luminance, glare and contrast in the workplace need to be balanced with other parts of the visual field because visual perception by the visual cortex depends on information from both central and peripheral vision.

Visual performance depends on the number of aroused cells of the retina followed by the number of neural synapses leading to the brain visual cortex. The more cells are aroused the higher visual performance can be; unfortunately this is not a clear-cut correlation, since performance also depends on the general intelligence of the individual. Performance is related to other mental processes such as cognitive function, memory, concentration and learning. The results of visual performance testing under certain lighting conditions should not be interpreted as only a function of lighting, ignoring other personal attributes. The level of lighting can only partially improve or impair brain processes.

Eye defects

The most common eye defects which may cause complaints during visual work are hyperopia, myopia, presbyopia, astigmatismus and convergence insufficiency.

Very few people have perfect vision. The distribution of refraction status in the population follows a Gaussian distribution; about 65% of the population have a defect which would require glasses to correct, but less than 65% actually wear glasses.

Some refractive errors are so slight or inapparent that a person may not be aware of the consequences (asthenopia and discomfort). Many people compensate for their visual defects by adjusting their viewing distance, tilting their heads or looking for the best working position. This can cause other effects such as headache, neck and back pain.

If the refractive power of the eye is too weak (because the eyeball is too short) to focus on near distance objects, incoming light rays are refracted behind the retina; the resulting effect is not a point but a patch, causing blurred vision. This defect can be partially improved by constant accommodation but this leads to eye strain and headache. People with slight hyperopia often compensate by moving visual objects away from their eyes. However, only proper correction with glasses or contact lenses will normalize the focus and visual acuity and remove eye disturbance.

If the refractive power of the eye is too strong (the eyeball is too long) to focus on distant objects, the incoming light rays are refracted short of the retina, resulting in a patch, not point focus and blurred vision. Accommodation cannot help to improve visual acuity; on the contrary it makes it worse. Partially myopic people help focus by simply bringing an object closer to the eyes. Only proper spectacles or contact lenses can correct this refractive error.

Loss of elasticity of lens fibres due to ageing prevents accommodation, thus limiting focusing light rays. Accommodation can improve visual acuity, but continual strengthening of accommodating muscles causes eye strain. The best correction is near-distance or multifocal glasses.

The cornea contributes about two-thirds of the eye's total optical power, with the other third contributed by the lens. Defects in corneal shapes distort the retinal imagination. Almost all eyes have some degree of astigmatism (the cornea is never perfectly shaped), but only a few people have this defect severely enough to interfere with correct perception. Astigmatism can be corrected with a special lens.

Convergence inefficiency may affect 20–30% of the population and is frequently associated with asthenopic symptoms of close work, especially blurring, double vision and headache as a consequence of the extra effort needed to make the eyes converge for viewing near objects during prolonged close visual work.

Other human factors

Gender

Many epidemiological studies have concluded that women report more symptoms of SBS than men; but few acceptable explanations have been found. Women work more frequently as typists performing data entry, so their long-term visual work tends to be more boring and less creative. Women often spend more working time with computers even though they prefer human–human rather than human–computer relationships [2]. Women physiologically have a larger pupil diameter (to control the amount of light entering the eye) so glare, for example, causes more eye discomfort than it does for men.

Age

Physiological changes in eye structures and in visual processes occur early relative to changes in other organs, with obvious consequences. Presbyopia (decreasing visual acuity) begins around 40 years of age. All visual processes such as accommodation and adaptation occur more slowly, lighting, colour and contrast sensitivity decrease, while eye symptoms resulting from visual work can be more frequent. However, if people in this age group wear corrective glasses complaints need be no more common than among younger people.

Type of job

Job stress and job satisfaction are probably more important causes of symptoms than physical and chemical factors in the indoor environment. The proportion of work time spent in front of computers is very important for both the number of eye symptoms and satisfaction with lighting conditions. Computers are considered an important source of visual problems, but in fact it is not the computer itself but the work with it that can be the cause of complaints. Different work from the conventional pen and paper and the necessity of learning new skills can trigger job stress, which creates an increasing number of complaints.

This does not mean that computer users need lighting conditions that are different from those for conventional office work. Work with VDUs emphasizes both weakness in lighting design and inapparent eye abnormalities that need no correction under conventional circumstances. Good ergonomic design of furniture (especially desks and chairs) and computer skills training can be more helpful than special lighting.

A high level of self-management, creativity and job responsibility has been related to satisfaction with lighting; on the other hand, boredom, submissiveness and monotonous jobs are correlated with increasing complaints about lighting. Clerks, typists and full-time VDU terminal operators complain more about lighting than managers and professionals.

HEALTH CONSEQUENCES OF INAPPROPRIATE LIGHT CONDITIONS

To assess properly the impact of light and lighting on human health disturbance, we must first define health. Health defined in WHO terms as well-being in the physical, mental and sociological sense seems to be ideal for assessing the health consequences of lighting. Although we have not described the set of clinical criteria clearly and the causes are not defined accurately, symptoms of SBS related to lighting and visual work can be categorized into the following groups: discomfort, asthenopia or general symptoms:

- discomfort (eye irritation): dry eyes, burning, redness, itching;
- asthenopia (eye strain): blurred vision, double vision, after-images, transient myopia, decreased visual performance, eye fatigue;
- general symptoms (non-visual): mental disorders and psychiatric illnesses, immune and hormone imbalances.

Eye symptoms following long-term visual work are caused by both inappropriate lighting and disorders of eye systems. General symptoms are strongly influenced by lack of daylight.

Eye discomfort

Redness, itching, burning or tears (eye irritation) are symptoms rarely caused by lighting conditions only; they are more influenced by dry air, high temperature, dust particles, chemicals and allergens in indoor air, via neural (*nervus trigeminus*) or mechanical irritation of the eyelids (frequent blinking).

Contact lens wearers can be more vulnerable to the consequences of visual work. The protecting tear film becomes thinner, so dry air, dust, particles, fibres, bacteria and viruses can easily irritate the eyes.

Asthenopia

Symptoms of asthenopia (eye strain) arise from the extreme muscular effort of eye structures during prolonged close visual work. Inappropriate lighting levels and individual eye deviation can lead to severe asthenopia.

Blurred and double vision

Differences in either the luminance of objects or surfaces in the visual field or rapid changes in contrast result in increased frequency of processes of adaptation in the retina, and a reduction in this capacity. Eyes adapted to inactual luminance became strained.

Too strong or too weak a contrast results in decreased visual acuity and the strength of accommodation mechanisms can cause symptoms of asthenopia.

The type of glasses can be important. Reading glasses are excellent for near work but can cause asthenopia during visual work when entering data or typing on a computer. Eye working distance varies for a screen, keyboard or documents, and the focal point can often be outside the range glass correction.

After-images

Work with fixed patterns and colours in the visual field can cause colour or brightness after-images. After viewing a surface of one colour for a period of time and shifting to a neutral coloured surface the complementary colour can be seen. For example, red surfaces create green after-images. A view of a very bright surface for a period of time is followed by a bright after-image after visual shift to a matt surface. These asthenopia symptoms are a consequence of the time shift of retinal adaptation.

Transient myopia

The spectral distribution of light when moved to long wavelengths (red, orange) together with low illuminance can cause transient myopia, especially for normal sighted young people. Temporary spasms of the lens muscle can cause transient myopia after a long period of close visual work under inappropriate lighting. However, there is no medical evidence that the transient myopia or spasms of tired eye muscles cause permanent defects.

People with myopia or presbyopia can work without complaints under lighting in the long wavelength spectrum because this has some advantages for their refractional errors.

Eye fatigue

Fatigue of any part of the human body is physiological prevention against its impairment. Continuing activity in spite of varying signs of fatigue can cause physiopsychic disturbance, even an injury. The term visual fatigue has been frequently used to define the consequences of visual performance, but pure visual fatigue is not measurable and no clear symptoms have been found. The visual apparatus is characterized by a number of flexible elements which can work independently of lighting condition and time period. The retina is free of fatigue. The process of accommodation is partially dependent on the work of muscles, so one measurable sign of visual fatigue might be a shift of the nearest focal point and a transient worsening of visual acuity.

Unconscious higher accommodating compensates for tiredness of the eyes; but if visual work continues without a break it can lead to asthenopia.

Assessment of the fatigue level can also be measured by the increasing number of errors made during a visual task. However, these findings are also influenced by general fatigue from both mental tension or boredom related to the work itself.

GENERAL SYMPTOMS (NON-VISUAL)

Mental disorders and psychiatric illness

These symptoms and diseases are predominantly caused by daylight deficiency but can affect the number and frequency of SBS symptoms.

Behavioural disorders

Hyperactivity, neurotismus and gaps in concentration and learning have been found among pupils and students spending time in windowless classrooms without daylight.

Syndrome seasonal depression

Symptoms arising from lack of daylight during the autumn and winter months include decreased mental and physical activity, increased food consumption and increased body weight, sleepiness and tiredness. Extreme fatigue can restrict normal daily activity during work, social and family life. The symptoms are basically caused by disturbance in the regular production of melatonin controlling other hormones. The rhythms of body functions of women are biologically more dependent on them than men are.

Psychiatric illness

Manic depressive psychosis and other types of depression, narcolepsia and sleeping disturbances can be caused by the lack of control of biological functions on the cell level. The role of melatonin and lighting control may be very important.

Immune and hormone systems disbalance

Immunity is strongly related to neurological processes. Melatonin production is one factor that influences the efficiency of the immune system because of the increasing number of natural antigens stimulating the immune response. Decreased immunity can cause a higher prevalence of infectious and other diseases. Loss of synchronicity in tissues and cells could result in abnormal mitic division of cells leading to cancer, especially cancers of the female reproductive system.

Melatonin due to light variance influences the menstrual cycle. Female sex hormones are released regularly since they depend on biochronological rhythms. Lack of daylight can cause disturbance in sexual function, mood and activity. Male sex hormones are not regularly released so the impact of melatonin on sexual function is not so obvious.

A red or orange colour increases sexual activity in male birds. The influence on men could be similarly based on the psychological and physiological attributes of colours.

Remedies

Two methods are available to improve health complaints and dissatisfaction arising from visual work: first, prevention at the stage of planning and design; and second, remedies to buildings in use.

LIGHTING DESIGN

The principles of lighting planning are described in many publications on architecture, light technology and construction, and a brief summary of related issues is given here. Lighting design should respect the size, height and depth of a room, the number of desks or workplaces, the type of work, use of VDUs, colour and surface quality and the geographical orientation of the building (especially windows). The choice of the main components of lighting design (both sources and system) strongly depend on these variables.

The quantity and quality of artificial lighting can be calculated according to many different methods; but not all of these methods take into account every criterion described above. Unfortunately daylighting is usually evaluated without considering the quantity and quality of light entering buildings. A more precise set of criteria for daylighting should be considered:

- illuminance in relation to the visual task (measured as minimum illuminance with a cloudy sky);
- illuminance distribution in the different parts of the room, including workplaces;
- direct glare from the sun and thermal discomfort from sunshine.

Lighting in existing buildings

Before attempting any remedies to the lighting environment as a response to health or environmental complaints from the occupants, a comprehensive survey to search for probable causes, is strongly recommended. Managers of buildings or companies where occupants complain about the lighting frequently prefer to measure illuminance only and to compare the readings to norms and guidelines; but assessment of lighting leading to remedies is more complex, takes more time and is possibly more expensive, but the results will benefit both occupants and manager. Improved satisfaction leads to better work performance without increasing job stress.

The optimal survey of the lighting environment includes the following essential steps:

- a preliminary review of the building to establish a schedule of measurement;
- measurement of lighting parameters (at least illuminance, luminance and luminance variation);
- measurement of indoor air quality (at least temperature, humidity and ventilation rate);
- a questionnaire for occupants about their perception of indoor environment and health complaints.

Recommended steps include physiological and biochemical examinations. The biochemical and psychophysiological markers of three sorts of general strain are summarized in Table 3.4. There are important differences among physical, mental and emotional stress to improve conditions which need it. Biochemical markers measured from blood samples are related to acute strain, while those from urine arise from long-term strain. An ophthalmological examination (for proper correction of refractional errors, treatment of eye diseases or allergy) is also recommended.

The measurement of lighting parameters is briefly described in Part II and is detailed, for example, in the CIBS *Code for Interior Lighting* [3]. WHO has adopted set recommendations for both the occupational and domestic environment from the CIBS *E Code*. The norms, guidelines or recommendations are very similar in other countries, and the European norm ISO 8995 describes the demands of appropriate lighting and the ergonomics of vision. A summary of current recommended values for office work is given here for direction, but the user should be familiar with the main documents mentioned above:

- Ratio day lighting illuminance: artificial lighting illuminance a minimum 1:5;
- Iluminance: 300–750 lx (1000 lx for plant or windowless offices);
- Luminance variations in workplace: 1:3:10;
- Colour rendering index: 80–90;
- Chromatic temperature/colour: 3300–5300 K;
- Shadowness : vector/scalar ratio: 2.0–1.5.

The general questionnaire on symptoms of SBS and environmental complaints completed with a special lighting questionnaire should be used to evaluate health and environmental complaints. The lighting questionnaire should include detailed questions about general satisfaction with lighting, decor and colours, sources of glare, flickering, contrast, colour appearance and control over lighting (both artificial and natural). General data about age, sex, type of job, time spent in the office and hours of working with a computer are also necessary.

Evaluation of data needs not only good statistical practice but also

Table 3.4 Biochemical and psychophysiological parameters of strain (adapted by Buscein [4])

Parameter	Category of strain			Other specifications
	Physical	Mental	Emotional	
Heart rate	**	*	*	– Adaptation * Effort, strength, concentration, fatigue
Breath rate	*			
Finger pulse		*	**	
Systolic blood pressure	**			
Diastolic blood pressure	*	*		
Eye blink frequency	*	*	**	fatigue
Pupillary dilation		*	**	
Core temperature	**			Circadian rhythm
Finger temperature			*	
Adrenaline		**	*	Effort
Noradrenaline	**		*	Muscular strength
Cortisol		*	**	
Free fatty acids			**	
Sodium/ potassium ratio	*		–	

* increased parameter is marker of increased strain during visual work
– increased parameter is marker of decreased strain during visual work

knowledge about the specific situation in the building where the survey is carried out.

If 80% of people in the building do not have complaints this is taken as satisfactory. The percentage of complaining occupants includes both people with higher sensitivity to the surveyed factor and people who are constantly complaining about something.

The best confirmation that the proper cause of dissatisfaction has been diagnosed and corrected is a repetition of the same steps as before, no sooner than three months after the remedies have been put in place.

Other recommendations related to lighting and visual work

- A windowless building should only be specified for limited, special types of work.
- Direct glare from windows should be avoided, by providing screening against direct sun or sky reflected light.
- Reflected glare should be minimized with matt surfaces and indirect lighting.
- Task lighting switches and window blinds should give control over light.
- There should be regular checks on air quality, temperature, humidity, ventilation and general cleanliness.
- Regular cleaning – maintenance and replacement of lamps and covers is recommended.
- Fluorescent tubes with the same chromatic temperature should be used – never mix warm and cold in one room.
- There should be proper management of full-time work with VDUs: frequent shorter breaks are more useful than only one long break; activity should be changed as often as possible, and the workplace should have a good ergonomic design.
- An assigned person should be responsible for discussion about the individual environmental and health complaints of occupants and for solutions regarding individual needs as well as common demands.
- For those complaining of permanent visual discomfort or strain proper glasses for the visual task distance should be supplied, and they should be educated about lifestyle improvements, including adequate sleep, stopping smoking, treatment of current diseases such as allergies, migraine, spinal disorders and all chronic eye diseases.

Designers create lighting for the 'average' population. However, whatever the quality of lighting design, it cannot satisfy all occupants. Managers and supervisors should try to adjust lighting to each individual whenever possible.

DISCUSSION

Few epidemiological studies focus on lighting as a cause of SBS. Evidence about light impact on humans has been obtained mainly under experimental conditions rather than from field studies, and contradictory results can be explained by the differences between methods and conditions.

Field surveys use both questionnaires and technical measurements. While questions enquire about health and environmental complaints over long time periods (weekly, monthly, or annually), measurements of lighting are usually taken during the short period of a working day.

Other explanations for differences in perception of lighting have focused on the individual attitudes of occupants.

Wearers of glasses may be a more sensitive group of the population. Hedge *et al.* described glasses as a variable that influenced the personal symptom index more than gender. However, they concluded that those wearing glasses did not complain more than other occupants about lighting [4]. There is no general suggestion that wearing glasses is strongly related to eye discomfort, but improper correction, rather then inappropriate lighting itself, can cause eye complaints [5].

The majority of studies on SBS conclude that women report more environmental and health complaints. Norback explained this as due to the higher prevalence of allergies (especially to nickel) and of frequent infections among women [6]. Hedge *et al.* concluded that there were no differences in allergy prevalence between men and women [4].

We discovered in our office lighting study [7] that women were more satisfied with general lighting in an office than men. We suggest that their dissatisfaction/satisfaction depends on the general lighting appearance in an office. Men concentrate more on details in the visual field and are less critical about general lighting, decor and colour but more critical about contrast and luminance in their workplace.

Laboratory tests conclude that the older population requires a higher lighting level, physiologically related to decreasing sensitivity to light, but the opposite conclusion has not been reached. Do younger people need less lighting? This is possibly true because children are able to play and read under poor lighting without suffering any eye complaints or fatigue. On the other hand, Lindner found that younger people need more light than older people; however, differences between groups decreased with the increasing difficulty of the visual task [8]. We also found a higher proportion of people under 40 complaining at the same lighting levels as people over 40 [7].

Adaptation of retina cells to low levels of luminance allows visual processes to function without discomfort; but if the work requires quicker visual performance, then the low illuminance can be reported as inappropriate [9]. Satisfactory time to adapt, better ability to concentrate and better professional skills are a good compensation for the slowing down of the physiological processes of vision during ageing. The retinas of young people physiologically react immediately to changes in luminance; but young people are also more aroused and disturbed by other environmental and social factors. They become mentally tired more quickly, which results in visual fatigue and possible complaints about lighting.

A compromise suggestion could be that younger people complain more without real health complaints, while older people complain less, even though they have real symptoms. Generally, the type of work and the visual task are more important than age in lighting design.

The distance from windows is often suggested to influence both the number and frequency of SBS symptoms. High levels of stress hormones due to greater distance from windows were found by Hollwich [1], but more comprehensive studies have failed to confirm this strong relationship between distance from windows and eye-related SBS symptoms [10].

Boubekri *et al.* concluded from comparison experiments in simulated light environments and a survey in real conditions that window size had no significant effect on mood responses. The significant changes due to variation in sunlight penetration showed opposite results in the two environments. While volunteers in simulated environments found sunlight to be relaxing, people in the office found sunlight exciting [11].

CONCLUSION

Light is perceived unconsciously; but its impact is stronger than people are aware of. If lighting is satisfactory it is not usually noticed; but inappropriate lighting influences well-being. We should agree with Herakleitos that 'the invisible harmony is more effective than the visible one'.

General complaints on lighting can be a first indicator of dissatisfaction based on causes far removed from actual lighting conditions – the work burden, poor human relationships or a low salary. Changes of lighting, which are often costly, must be considered from all possible viewpoints, including consideration of less expensive and easier alternatives.

Lighting technology tends to use high illuminance sources in interiors; but this has no linear correlation with satisfaction. The sun as a source of light has one great advantage which can never be achieved by any artificial source in buildings : great distance from the visual field followed by balance in lighting distribution, natural variance in intensity and spectral composition which benefits eye perception. Since daylight intensity has been suggested as the pattern of ideal lighting, artificial lighting is suggested to be, at best, a less appropriate substitute.

Humans as biological species developed for a long time period under natural daylight conditions; it should not therefore be expected that artificial light can be fully sufficient. Artificial lighting can be made satisfactory for human perception as a compromise, facilitating modern life, together with its other 'unnatural' components. Artificial lighting is necessary and has benefits; but it will never achieve the same quality as daylight.

REFERENCES

1. Hollwich, F. (1980) The Influence of Light via Eye on Hormones in Man and Animal, *Proceedings of the CIE Symposium*, Berlin, pp. 182–94.
2. Durndell, A. (1992) The gender gap in IT in Britain, in *Work with Display Units 92*, North-Holland, Amsterdam, pp. 527–31.
3. Chartered Institution of Building Services Engineers (1988) *Design Guide Data*, Vol. A., London.
4. Hedge, A., Erickson, W.A. and Rubin, G. (1995) Individual and occupational correlates of the sick building syndrome, *Indoor Air*, **5**, 20–1.
5. Cakir, A.E. (1991) *Light and Health: An Investigation on State-of-the-art and Future Prospects of Lighting Technology in German Office Environments*, ERGONOMIC, Berlin.
6. Norback, D., Michel, L. and Widstrom, J. (1990) Indoor air quality and personal factors related to the sick building syndrome. *Scandinavian Journal of Work and Environmental Health*, **16**, 121–8.
7. Drahonovska, H., Burge, P.S., Smith, N.A. and Calvert, I.A. (unpublished) Lighting office study.
8. Lindner, H., Palm, K. and Schlote, H.W. (1987) Subjective lighting needs of the partially sighted, *Proceedings of the CIE Symposium*, Genoa, 102–3.
9. Van Ireland, J.F.A.A. (1967) *Two Thousand Dutch Office Workers Evaluating Lighting*, Research Institute for Environmental Hygiene, 283, Delft.
10. Robertson, A.S., McInnes, M., Glass, D., Dalton, G. and Burge, P.S. (1989) Building sickness: are symptoms related to the office lighting? *Annals of Occupational Hygiene*, **33** (1), 47–59.
11. Boubekri, M. and Boyer, L.L. (1995) A comparative study of building occupant response to luminous displays in real and simulated indoor environments, *Indoor Environment*, **4** (2), 113–20.

CHAPTER 4

Medical aspects

Keith Eaton

The descriptive title of the syndrome would appear to suggest, correctly as the rest of the text shows, that it is the building that is ill. As physicians we are rarely referred buildings as patients, but commonly to the people who dwell or work in them, and it is from their reports that our diagnoses and proposed treatment for the building must be generated. The management of the building itself lies outside the medical remit of this chapter. However, we must consider the nature and extent of complaints in affected subjects, relevant investigations, differential diagnosis and the management of the problem, which will invariably stray beyond the usual remit of the physician.

When I was a medical student in the 1950s the condition was not mentioned, and indeed was not raised as an entity during my post-graduate training. Even in the 1960s it was being discussed solely at research meetings and did not significantly enter the public domain until the 1970s: today most people have heard of it, even if they are generally not equipped to suspect the nature of the diagnosis. A novel diagnostic entity may be new or unsuspected, and increased reporting may result from either an increase in awareness or a genuine increase in the number of cases. Epidemiological studies, properly carried out, can shed light on the nature of the illness, and can serve to confirm data from other sources about theories of the cause. It is regrettable, but hardly surprising, that no formal studies have as yet been done as to the incidence of sick building syndrome in the population as a whole. In a computer search no literature references to this topic were identified.

SBS

The author's observations, based on 35 years of practice, are perhaps widely shared and would be confirmed by others working in the field.

- Our medical ancestors were keen and experienced observers. No reports of SBS have been found in the UK which pre-date the 1939–45 war.

- Affected subjects are normally improved or even symptom free at home and/or on holiday.
- They normally have experience of work in other buildings where they have not been adversely affected.
- The buildings in which they are troubled are normally new rather than old, and almost invariably have comprehensive air conditioning which controls the indoor climate: to secure this control openable windows are not fitted.
- Transfer of patients from these buildings to other less controlled environments, but doing the same work, will normally reduce or abolish symptoms, in the absence of other treatments.

The above observations suggest that the cause of the problem lies in the work environment of particular buildings.

It may seem both unnecessary and undesirable for the medical chapter to reiterate observations which must be found extensively elsewhere in the text: the author would, however, contend that it is part of the remit of the clinician to extend his or her observations beyond the simple clinical management of patients, as by no other means can appropriate treatment and preventative strategies be evolved; and furthermore this condition may well only be suspected when there is an input from management and/or employee representation at the workplace. In current UK medical practice the first point of medical contact for most employees will lie with their general practitioners, and under most circumstances employees will be registered with a number of different doctors. The individual symptoms do not form a particularly striking pattern in themselves, and the author suspects that SBS as a diagnosis is missed more often than it is picked up and treated.

An example may illustrate the problem. Some years ago a patient was referred for a second opinion to the author who, from prior experience, did suspect that SBS might be the correct diagnosis. The patient was found on enquiry to work for a small firm with under a dozen employees which occupied part of one floor of a modern air conditioned four-storey office block, each floor of which was similarly divided into small units. The patient was unable to trace another similarly troubled patient in her own firm and there was no point of contact with any of the other firms in the building; the company responsible for services management was unwilling to investigate further unless there was clear evidence of a generalized problem. Thus no progress could be made. The patient was advised to change her place of employment: having done so her health improved. Some years later there is still an unconfirmed suspicion that firms in this building come and go rapidly, and have a high staff turnover. Would the outcome have been different with a greater level of public awareness, assuming that the suspected diagnosis was in fact the correct one?

THE CLINICAL SYNDROME

While the question of an input from employees and/or employers remains highly desirable this is largely an issue for the future: under current conditions of practice it is the symptomatic individual reporting to his or her general practitioner with whom we shall mainly be concerned. We must therefore consider the nature of the symptom complex. At once we must realize that there is no specific group of symptoms or signs which occur in this condition which cannot arise from other causes. The complaints may be divided into organ symptoms in the usual way. (The author contends that such divisions are artificial, and tend to hold back progress in medicine since much environmental disease is *not* organ specific, and studying it in a single organ system involves ignoring or missing symptoms arising elsewhere, with a danger of making an incorrect diagnosis. However, much specialist medical advice in this country is organized on the basis of single-organ-system specialisms.)

- Neuro-psychological: patients chiefly complain of fatigue, malaise, impaired memory and concentration, visual disturbances and headaches. Episodes of collapse have been seen.
- Respiratory tract: nasal symptoms of stuffy nose, sneeze, discharge and blockage, cough and occasional wheeze predominate. Some patients have eustachian catarrh, causing mild catarrhal deafness; tinnitus may also occur.
- Dermatological: dry skin, and skin rashes, including urticaria, may be seen.
- Musculoskeletal: muscle and joint pains and fibrositis are occasionally seen, but are not usually a major complaint.

As will be seen none of the above are specific to SBS and this constellation may result in patients being referred simultaneously for several independent specialist opinions. If there is an appreciation that all the symptoms are linked this may well result in a psychiatric referral, particularly if it is appreciated that they appear to be work related.

CLINICAL INVESTIGATION

Appropriate channels of referral include allergists, of whom there are few in the UK. Therefore in practical terms referral may be made to specialists in industrial medicine, occupational health, departments of environmental medicine, or consultation may be made with the Health and Safety Executive. Investigation and management of these patients will necessitate investigations at the workplace, and therefore a multidisciplinary team will need to be available for a satisfactory management to be achieved. It is desirable that one physician should be in overall control of the clinical side, and should see all the patients, and be in a position to coordinate investigations into possible unreported problems in other workers on the same site.

The first step, as always, should be a comprehensive medical history. While this should be automatically true for any medical consultation, in practice many doctors have become used to curtailing this step, and indeed many have not taken a *full* medical history since they were under-graduate students: the GP is working within the constraints of brief con-sultation times to assess a current complaint and most specialists in organ-related disciplines deal with complaints which may not require consideration of problems in other organ systems for which some other carer is responsible.

Here, as we have seen, the onset may be insidious and the symptoms diffuse; an adequate history is therefore essential. Many sufferers have a family history of allergy or allergy-related illness. Other illnesses may enter the differential diagnosis. A personal history of allergy may be pre-sent. Often this may have been in childhood, and the allergy may not have been active for many years, but needs to be sought: the allergy may not have been diagnosed at the time. Other causes of chronic ill health may include unusual problems such as Lyme disease, toxoplasmosis, brucellosis and the like, which will only be suspected if history taking is meticulous. A patient with a chronic fatigue syndrome may become ill coincidentally with an outbreak of SBS. Such a patient may or may not be made worse by the factors present in the building, but the primary cause of illness is different, and needs to be recognized.

When it has been noted the timing of symptoms in SBS can be helpful, although we must remember that patients are not trained observers, and the timing may need to be elicited by enquiry at a subsequent date. Nevertheless many patients do notice that their symptoms occur in rela-tion to work: they are better on holiday and at weekends, and worse dur-ing the working week. The pattern in the week is variable. For some Monday morning is bad, followed by a slight improvement on Tuesday and then gradual decline over the rest of the week. For others the effects of the working week are progressive and cumulative. The afternoons, especially just after lunch, are usually worse than the mornings. Not infrequently there are clusters of patient involvement. Those working in a particular area may be worse affected than those in other locations in the same building. Some areas may escape altogether. Some individuals may suspect that particular aspects of work make them ill, and these complaints should always receive careful assessment.

Psychological health should also be part of our assessment. This is a complex area. Often sufferers have had no adverse psychological pre-morbid factors, but not infrequently individuals who adversely respond to stress are afflicted in SBS outbreaks. This raises the question of the psy-chological component in causation, which will be further considered below. A good psychiatric component of history taking is clearly essential. Hobby and out of work activities, and factors in home life can contribute to illness and may in individual cases masquerade as SBS; again these should be part of a competent history taking.

It is to be expected that the above list will be incomplete, as the lines of questioning that will develop in individual instances can only be perceived at the time and cannot always be laid down in advance as a pro forma. The clinician must always be alert to areas which demand exploration. To acquire such a history from a patient will normally in experienced hands take about an hour per person.

Physical examination of the patient is normally a standard part of any medical consultation. However, in SBS this is usually remarkable for what it does not reveal, rather than what it does. However, if this step is omitted other diseases which may exist in parallel will be missed, and a physical examination should therefore be done. It is not appropriate here to reiterate standard medical texts in relation to the merits of physical examination.

Specific investigations are, however, of some value. Where respiratory symptoms are present respiratory function tests should be done, and should include peak expiratory flow and the forced expiratory volume in one second and forced vital capacity using devices such as the vitalograph. Testing for allergic status may be of use, as many sufferers do tend to be allergic, and specific occupational allergens were a constant feature in early outbreaks of SBS [1]. There were mould spores, generated by the type of air conditioning then currently in use, where humidification of circulated air was achieved by blowing the air over baffle plates sprayed with recirculated water: the consequence was mould growth on the baffle plates, and contamination of the air flow by mould spores from these colonies. Because of fears about Legionnaires disease, rather than SBS, this type of air conditioning apparatus is now obsolescent, and in any event scrupulous maintenance should avoid the problem.

There are a number of ways of determining classic atopic allergic status, but in experienced hands prick skin testing will give good results in ten minutes. For those unfamiliar with it reference should be made to specialist texts [2]. Reading should be supplemented with practical experience of attending a clinic where the technique is in regular use. Some experienced specialists may choose to use intradermal tests, but generally prick tests will be preferred for their greater convenience. A prick test programme can monitor atopic status (the ability to mount positive skin tests to common inhalant allergens [3]). Positive skin tests without a history incriminating the specific allergen do not indicate an active allergy – merely atopic status. A prick test programme should include a negative and positive control solution, house dust, house dust mite, common pollens such as a tree mix and grass pollen, animals to which the patient has been exposed and a mould spore mix. Experienced workers may wish to use a specific range of individual mould spores. The author would add a specially made extract of dust sampled from areas within the affected building where the number of affected cases is high.

Laboratory investigations may be of value particularly where adverse reactions to volatile organic compounds (VOCs) are thought to play a major part in the outbreak [4]. Generally these will be detected at excessive levels if consulting engineers or specialisms other than medicine are involved to take and analyse air samples within the affected structure at appropriate sites. However, direct measurement in patients is also of value.

First the objection may logically be made that the mere presence of VOCs in the air of a workplace does not prove that these enter the patient's body, or that in so doing they are associated with illness. Secondly, an employer or the owners of the building may be averse to the suggestion that their building is causing health problems, and patient measurements may be required to convince them. Thirdly, the question may be taken a stage further when it is the employees, perhaps via a union or staff association, who are making the complaints, and the employers or building owners may refuse permission for measurements to be made within the building; here sampling of patients is the only option. Finally, in the event of any legal procedures proof of a logical cause and effect may well entail demonstrating the entire sequence of events if a court is to be satisfied.

When it comes to laboratory tests on patients it has to be said that those required are not simple routine methods which all physicians and GPs are likely to have done but are specialist tests which most laboratories will not offer as a routine. However, although not all laboratories will offer appropriate procedures as a matter of course a number of good laboratories can perform most of those required if a well-reasoned request is put to them. The author has most experience of working with a private sector laboratory, the Biolab Medical Unit, The Stone House, 9 Weymouth Street, London W1N 3FF, Telephone 0171 636 5859/5905, who have pioneered testing in this area and would be happy to advise other laboratories on techniques and methodology. Their standard procedure is currently a toxic effects screen, which measures firstly glutathione-S-transferase (a test not generally performed elsewhere) together with sensitive liver enzyme tests, STALT, AST (transaminase) GT (transpetidase) together with S-nucleotidase. With the exception of the S-nucleotidase other good laboratories can offer these tests. A urinary D-glucaric acid excretion completes the package. This test again can be performed by other laboratories.

This toxic effects screen demonstrates that xenobiotic chemicals are placing bodily biotransformation (detoxification) mechanisms under stress, but of course is not a direct measurement of specific effects from individual chemical compounds. Evidence of this is hard to come by, but it is known that VOCs are detoxified in the liver [4] and in the absence of other known adverse chemicals in the patient's environment is strong evidence of a work-related problem, particularly if measurement of VOCs in the workplace has been performed. It should be noted that standard automated liver function tests are not sufficiently discriminating to be expected to show positive abnormalities in most subjects. Direct measurements of some VOCs may be made at Biolab using a pesticide screen and/or a detergent screen. So far this procedure is not available elsewhere. The same laboratory is currently developing a formaldehyde sensitivity test which is not available at the time of writing. If further confirmation of the work relationship is required repeated measurements performed in relation to a patient's known work attendance may show a clear and diagnostic relationship. This is illustrated in Fig. 4.1 where the subject was made ill by exposure to VOCs from mineral oils in a machine shop.

MANAGEMENT

Much of the current approach to conventional medicine involves identification of symptom complexes which can be controlled with the drugs which suppress the ensuing symptoms. Such an approach tends to be ongoing, as the cause of the symptoms may persist, and may not be explored or resolved.

An environmental approach to medicine, which is always an appropriate strategy in SBS, may obviously make appropriate use of drugs; but in

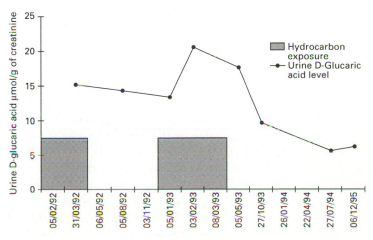

Fig. 4.1 Hydrocarbon exposure – male age 56.

addition attempts to identify causes which then enable alternative strategies to be deployed. These include avoidance and occasionally desensitization. In SBS it is chiefly the former which will be involved, and as the latter requires specialist training and clinic facilities it is not considered appropriate that it should receive detailed consideration in this text. It is, however, worth stating that using appropriate strategies there may be occasions when experienced specialists may wish to consider it as part of the treatment regime.

The most important factor, however, must be avoidance, and there are various ways this can be achieved. The simplest is when the clinician may advise the patient to change his or her place of employment, and may suggest that work is likely to be symptom free in an environment where ventilation is natural, and windows can be opened, rather than in a modern complex building where air entry is controlled and a modern heating and ventilating plant controls the environment. Such advice, while easy to give, should not necessarily be lightly considered, or indeed be the physician's first thought.

While this will perhaps remove symptoms for an individual patient it may mean that cause has not been established, and it may also mean that the patient may well in the future again be exposed to the same substances to which he or she is sensitized, and to which a further reaction will occur: since the patient cannot continue to change jobs it may well be better to adopt a more fundamental approach at the outset. In today's society changing jobs is not necessarily easy, and a patient afflicted by SBS may not always find it easy to obtain a new job. A change of job may involve undertaking new work for which he or she has not acquired trained skills; the patient may of course have trained skills which *are* of value in their employment where SBS is a problem. This may mean that the employee loses money as a result of downgrading and at the same time that the employer may lose a skilled operative whose replacement will take time to retrain, and also be at risk from acquiring a susceptibility to SBS. However, job changing does occasionally remain the best advice, particularly perhaps for patients with other pre-existing illnesses, or where earlier measures have not been successful in controlling the problem.

Having suggested that the route of redeployment is not the first approach in what order should we attempt to address the problems of the illness? The author would suggest that the first issue is one of information. The employer will, or should be, concerned, and it will probably be advisable for the physician to have meetings with the key personnel, preferably as part of a multidisciplinary approach to explain the nature of the problem, and the specific causes which have been identified, so that strategies for rectification can be adopted (see below). Secondly, the medical information available should be given to the workforce. Those who are clearly affected should be given full detailed information and counselling at one or more sessions by the physician. It may well be useful for this to be performed as a group, as support for other sufferers is known to be helpful in many chronic disease situations; but this should always be followed up with one to one sessions at which patients' individual problems can be properly addressed; follow up visits will almost certainly be required to assess progress and to ensure that appropriate action strategies are adopted. However, the information process should not stop at this point, as non-affected personnel may also have concerns: is it infectious? What are my chances of becoming ill? Are there things I can do which may help or hinder? Is anyone to blame? Such issues should be addressed. Equally information to the supposedly well population may reveal cases which have not previously been identified.

The minimum (but often sufficient) approach should be an informative letter from management, preferably jointly with the investigating physician, indicating channels of communication with medical advice for any medical queries which may arise. When a new attack of SBS has been confirmed a most valuable role can be performed by a properly trained and informed occupational nurse. Large employers may already have such an employee, but for others it might well be worth considering the recruitment of an industrial nurse, perhaps on a short-term contract of six to twelve months. From the viewpoint of a firm employing highly trained personnel who may be able to take their skills elsewhere, such an appointment may well be extremely cost effective and, of course, will enable staff to make informal contact when the physician may not be available. Smaller firms may be able to do this on a sessional basis: if so the nurse's availability should be circulated in writing to all employees.

The burden of this author's views on SBS will be seen to have thus far addressed only the physical aspects of the disease; but it would be inappropriate to assume that these will form the totality of the illness. For all individuals wellness involves an interaction between mind and body. We are all aware that the pain of severe injury can be ignored by soldiers in the heat of the battle by individuals who after the incident are subsequently incapacitated by their pain. At the opposite end of the scale those with pre-existing psychiatric morbidity may find that physical ailments which would be negligible to others become so intolerable as to produce a complete inability to function. It is a truism often repeated that a modern working environment is apt to be stressful. (The author considers that we tend to flatter ourselves in this respect: if in the Middle Ages you knew that the Black Death was in the next village the stress might well have been worse than knowing that your job is on the line in the 1990s.)

Whatever the degree of stress it must always be considered by the managing physician, as there are many literature references which detail the adverse effects of stress on the immune system [5]. These will affect the ability to mount a normal immune response to allergy, infective agents and carcinogens. It is not known whether the ability to metabolize xenobiotic chemicals will be similarly affected, but it is quite plausible that it might be, and significant perceptions of stress and psychiatric illness merit attention in their own right, regardless of whether or not other links can be substantiated. Most physicians have limited psychiatric and counselling skills and the decision to employ counselling and/or psychiatric drugs should be in the hands of those who possess appropriate training in this area. However, it is also essential that such a specialist must also be briefed as to the physical problems which are being addressed so that advice given is appropriate; and thus again the availability of a psychiatrist or psychologist as part of the multidisciplinary team is likely to be the best approach. One essential step in reducing stress is that management must be persuaded to secure the jobs of affected individuals while the problem is being dealt with, and to provide adequate reassurance to this effect. Again, as part of both management policy and stress reduction securing the cooperation of any workplace trade union and/or staff association should be regarded as essential. Most large unions do have medical advice available to them, at least centrally, and if a union is involved local organizers should be persuaded by the physician to ensure that the information reaches the union's medical adviser, and that his or her advice is sought and adopted.

So far this physician's approach has been entirely general and non-specific. The common first approach of most UK doctors is drug treatment; but in this case there is little to say. It is obvious that palliative remedies can be used, for example analgesics for headache, decongestants for nasal obstruction and the like. It should, however, be clear that none of these will address the cause of the condition, and that most employees will not be content to have to continue consuming levels of prescription medicines, possibly for years on end, to relieve work-related symptoms, particularly if they know that this would not be required if they worked elsewhere, and that many of their colleagues do not have to take them. It is logical, and better scientific sense, to address the cause of the problem.

Such an approach leads us away from what is generally regarded as the orthodox field of physicians treating patients: we must concern ourselves with the way heating and ventilation systems work, with humidity and air flow measurements, with areas of overcrowding and poor ventilation, with how the system is designed and maintained, with whether it may be generating or circulating allergens and with equipment and structures within the building which may generate VOCs.

Are lighting levels appropriate? Is smoking permitted and where? Can contaminated air re-enter the ventilation system and thus be recirculated? Few physicians have the skills to address such questions, which are always best answered by specialists who should be recruited to provide the appropriate solutions. Experienced doctors may be able to assist by explaining the significance of the findings, or by relating them to patients' illnesses, but essentially it is not the physician who at the end of the day treats the sick building. In the UK doctors can only treat animals as part of an ethically controlled collaboration with veterinary surgeons. In SBS the same principle should hold good, although the collaboration will be with experts such as consulting engineers.

The merits of the multidisciplinary approach may be illustrated by an example from the author's experience [6]. An episode of SBS occurred in a newly commissioned office building belonging to a multinational company: employees suffered skin and respiratory problems, and there were episodes of collapse. A multidisciplinary team investigated the outbreak and steps were taken to reduce the level of VOCs, to provide education, and maximize the efficiency of the ventilation. As a result there were no new cases, and the work environment became tolerable for all employees who remained: two were transferred, one to work abroad as part of a pre-existing agreement and one to another building within the company because of a pre-existing post viral fatigue syndrome. By chance at the same time as the team was called in the company was completing on the same site a building of identical size that was in essence a mirror image of the first. During the commissioning stages we advised on dust hygiene measures, on commissioning procedures for the ventilation system and on the selection of furniture and fittings which had been pre-offgassed to reduce VOC generation. The consequence of this was out of a thousand employees in the second building only one employee had minor symptoms, and these were so minor that he stated that but for the enquiry made by us he would not have reported them and did not consider himself in need of medical treatment.

SBS can be resolved with effort and goodwill on all sides. The medical input should be by experts as part of a multidisciplinary approach. With greater dissemination of information, in which we hope this book will assist, it could become largely an historical problem.

REFERENCES

1. Pickering, C.A.C., Moore, W.K.S., Lacey, J., Holford-Strevens, V.C. and Pepys, J. (1976) Investigations of a respiratory disease associated with an air conditioning system, *Clinical Allergy*, **6**, 109–18.
2. Eaton, K.K., Adams, A. and Duberley, J. (1982) *Allergy Therapeutics*, Balliere Tindall, London, pp. 48–55.
3. Mygind, N. (1986) *Essential Allergy*, Blackwell, Oxford, pp. 51–4.
4. Roe, D.A. (1991) Interactions of drugs with food and nutrients, in *Nutritional Biochemistry and Metabolism with Clinical Applications* (ed. M.C. Linder), Elsevier, New York, pp. 559–71.
5. Eaton, K.K. (1990) Stress and the immune system: the use of the term psychosomatic, *Newsletter of Society for Environmental Therapists*, **10**(2), 64–7.
6. Eaton, K.K. and Owen, B.J. (1993) Building related sicknesss: remedies for recovery, *Building Services: The CIBSE Journal*, **15**(3), 26–8.

CHAPTER 5

Psychological issues: a multifaceted problem, a multidimensional approach

Vyla Rollins and Gill-Helen Swift

Following an extensive review of the published research on SBS it seems that there is a concentration on the more tangible causes of the problem, such as indoor air quality (IAQ) and other pollutants; these tend to be the main areas of investigation. On the back of this maintenance and cleaning companies are creating a lot of business from changing and/or cleaning the air systems and interiors of organizations. These studies and companies are not always successful in either diagnosing the cause or eradicating the problem of SBS.

WHY A MULTIFACETED APPROACH IS NECESSARY

A reason for the lack of an accurate cause or 'cure' is that these investigations are only looking at part of the problem; a whole solution is therefore not able to be obtained. If the solution was purely physical, i.e. IAQ, then surely SBS would be a thing of the past and the cure simply a matter of taking measurements of the indoor air quality to identify the cause in order to recommend the appropriate course of action to eradicate the problem.

Failure to thoroughly investigate all facets, particularly psychological factors, such as job satisfaction, locus of control and work-related stress, may lead the investigator, whether academic or commercial, to attribute all symptoms to a physical source. SBS is more complex and not only includes problems with the physical environment but also psychosocial problems which could, if not recognized, not only exaggerate symptoms but possibly lead to susceptibility to symptoms through increased stress in working environments. Therefore not only is it necessary to measure and investigate environmental factors in the pursuit of eradicating SBS but also to measure and investigate the psychosocial problems which may also be contributing to an outbreak of SBS.

Inadequate holistic investigation of the problem may also lead to misdiagnosis. For example, it is suspected that many cases of SBS and neurotoxic disorders (NTD) are misdiagnosed as mass psychogenic illness (MPI) [1]. Not only is a multidimensional approach, both physical and psychological, necessary to eradicate the problem but also to ensure that the diagnosis is accurate in the first place.

For an accurate diagnosis of workplace-related illnesses the investigation must include not only a thorough industrial hygiene examination that documents levels of neurotoxins and other pollutants, but also building characteristics, the working environment, job demands and detailed psychosocial assessments particularly of work-related stress, social support, perceived control, management issues, organizational culture and change processes.

THE WORKPLACE-RELATED ILLNESS MODEL

The model in Fig. 5.1 shows not only the interlinking facets of SBS but also distinguishes the relationship of these facets to the other three types of workplace-related illnesses. At this point it is important to distinguish the

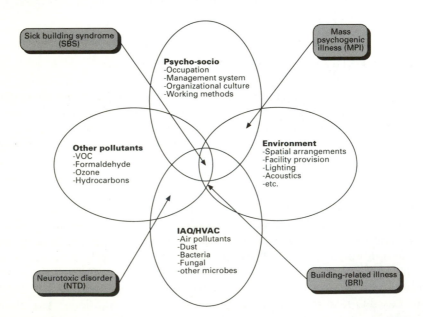

Fig. 5.1 The workplace-related illness model.

underlying differences of the four main types of working environment-related illness and to discuss the underlying features of assessment that are applicable to any investigation into workplace illness. This will not only clarify each type of problem, but will also avoid misdiagnosis of the problem so that an appropriate course of action can be followed. Although these disorders may have different causes all have sufficient superficial similarities, and many hygiene and building professionals may confuse them, thus producing misdiagnosis and inappropriate treatment. The characteristics that they tend to have in common are that a number of people working within proximity of each other exhibit similar physical and psychological symptoms that tend to be either provoked or be associated with an environmental incident such as an unidentified pollutant.

The four main types of workplace illness are:

- neurotoxic disorder (NTD);
- building-related illness (BRI);
- mass psychogenic illness (MPI);
- sick building syndrome (SBS).

DEFINITION AND DIAGNOSIS OF THE FOUR TYPES OF WORKPLACE-RELATED ILLNESS

NTD

When people are exposed to neurotoxic substances such as heavy metals and mixtures of organic solvents, symptoms of mood changes, motor and mental slowing, memory problems and problems with concentration can be seen [2]. NTD differs from SBS and BRI as both the physical symptoms and the psychological changes are pronounced. However, NTD may produce a similar psychological reaction to SBS, particularly if the solution to the problem is ignored or delayed. Exposure to levels of neurotoxins that approach or exceed government or public health standards, or a long history of chronic exposure to low-to-moderate levels, are generally responsible for NTD. NTD may directly affect the central nervous system. It is considered to be unrelated to gender and psychosocial variables like work-related stress, although symptom severity may be related to factors such as age and duration of exposure to the neurotoxin.

However, many individuals will show significantly high levels of psychological distress on psychological tests such as MMPI and the SCL-90. This is most probably due to exposure to the individual chemical, i.e. atypical post-traumatic stress disorder and not to the individual's personality or circumstances. It can therefore be said that there is a direct causal relationship between exposure of the physical variable, i.e. the neurotoxin, and the occurrence of the symptoms. NTD thus has a known aetiology which, with adequate investigation of physical measures of air quality, particularly if the measures are in excess of government or legislative figures, will lead to a NTD diagnosis.

BRI

Analogous with NTD, building-related illnesses have a known aetiology with specific symptoms and lab findings. Humidifier fever and Legionnaires disease are classed as building-related illnesses. Building-related asthma and allergic rhinitis may also be considered in this category, but they are more difficult to distinguish from asthma and mucal irritation symptoms reported in SBS [3, 4]. As with NTD, these disorders are not associated with psychosocial variables such as gender or work-related stress, and apparently unrelated to building characteristics like lighting [1]. BRI can be diagnosed with detailed and thorough investigation of the physical aspects of the building and the spread of symptoms.

MPI

MPI is defined by Colligan and Murphy [5] as 'the collective occurrence of a set of physical symptoms and related beliefs, in the absence of, an identifiable pathogen'. Outbreaks of MPI share common aspects with all the other workplace-related disorders – SBS, NTD and BRI.

Five predictors that accounted for more than a third of the variables of an outbreak of MPI have been identified [6]. These are:

• work intensity
• mental strain
• work/home problems
• education
• sex

in that order of importance. Psychological problems are therefore most likely to be the predisposing factors in an outbreak of MPI rather than a physical or building-related cause for NTD or BRI.

Individuals who develop MPI are psychologically vulnerable, usually as a result of high levels of stress within their working environments. Their level of psychological dysfunction is usually long standing and precedes the outbreak of MPI [1]. If no physical aetiology can be established, psychological assessment becomes an essential part of the investigation in such cases.

MPI differs from SBS as the symptoms do not usually remit when an individual suffering from MPI is removed from the building. The spread of MPI differs in that MPI symptoms tend to spread through social networks, unlike SBS, which tends to cluster in particular sections of the affected building and may be associated with particular groups. Thus, looking at the physical components, the psychological aspects and the clustering of symptoms allows a distinction to be made between a diagnosis of MPI and SBS.

SBS

SBS is recognized by the World Health Organization. SBS is suspected if a large number of a buildings occupants (usually 20% or more) experience symptoms that cause acute discomfort but for which no consistent aetiology can be established. Typically the people suffering from SBS work in proximity with each other and manifest similar somatic and psychological symptoms which tend to be triggered by or associated with an environmental event such as the emission of an unidentified pollutant, the perception of an unusual odour or contact with a toxic chemical. Relief from the symptoms is generally experienced when sufferers leave the affected building at the end of a working day, over the weekend or during a leave of absence. These symptoms then return once they are back in the affected building.

Five general symptoms are associated with SBS [7, 8]:

* mucus membrane irritation, which usually effects the eyes, nose and throat;
* neuropsychiatric disturbances, such as fatigue, headache, confusion and dizziness;
* skin disorders, for example itchiness, dryness and rashes;
* asthma-like symptoms, such as tight chest and difficulties in breathing; and
* unpleasant odour and taste sensations.

Figure 5.2 gives the frequency of symptoms found in one study [9]. Although these symptoms are found to be common in the general population the incidence is very much higher in occupants of specific buildings [10], particularly those buildings that are large, mechanically ventilated and have sealed structures with large areas of open plan.

The authors [9] also indicated that that symptoms were twice as frequent in buildings that were centrally or locally supplied with induction/fan coil units than in naturally ventilated buildings. Symptoms were increased substantially once an air supply was either chilled or humidified, in a building with centrally controlled systems managed by a property or facilities group.

However, most investigations fail to find a single chemical exposure that produces the symptomology in all cases of SBS. Whether in new, refitted or naturally ventilated buildings, the non-specific nature of SBS symptoms and the frequent failure to effect a speedy resolution of any IAQ (indoor air quality) problems has not only led to the scepticism

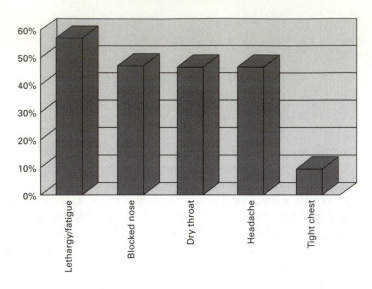

Fig. 5.2   Frequency of symptoms reported by Burge *et al.* [9].

about the existence of SBS, but also to the suspicion that sufferers from SBS are hysteric and thus displaying MPI.

Therefore by definition SBS symptoms cannot consistently be attributed to or directly linked with known toxins such as formaldehyde exceeding established standards or obvious bacterial and viral disease, all of which form NTD; and as it is not possible to attribute all symptoms to a single chemical toxin or disease then it may be appropriate to say that the building is somehow dysfunctional [1]. However, the occupants or the organizational culture may also be considered dysfunctional and it is this combination of inappropriate building and occupant mix which may exaggerate or prolong the problem.

The psychosocial connection

The initial cause of SBS symptoms may well arise from a physical source; however the psycho-social structure of the organization may have an effect on the continuing occurrence, and relief, of symptoms. The physical event that initiated the onset of symptoms, if recognized at the time, may have easily been solved or eradicated if acted upon immediately. However, if there is a delay in the response to the physical event that caused the onset of symptoms, then symptoms may not only be exacerbated by the psychosocial structure but may also affect the eradication of the SBS symptoms within a building.

If left undetected for sufficient time for the symptomology to become established, just tackling or investigating the physical cause may initially lead to a 'cure'; but unless the psychosocial problems are also dealt with the symptomology may reappear or never truly diminish.

Locus of control and learned helplessness

One major motive of competence is the need to be in control and not at the mercy of external forces [11]. The need for control is closely associated with the need to be independent of the controls and restrictions of others and to be able to dictate one's own actions, not to be dictated to or have actions determined by others [12]. When freedom or control are threatened people tend to react by reasserting their freedom, i.e. exhibiting SBS type symptoms; this is called psychological reactance.

The first experience of not being in control is likely to produce reactance; but with continued loss of control the reaction is likely to become learned helplessness, i.e. the continued expression of SBS symptoms. Learned helplessness, according to Seligman [13] is a learned reaction that no behaviour has any effect on the occurrence (or non-occurrence) of a particular event. The effects of learned helplessness are not easily changed; however, after many trails of positive reinforcement the association or reaction can be changed to a positive behaviour or reaction, thus eradicating the learned helplessness behaviour.

Locus of control and the working environment

On the whole a building's occupants tend not to feel that they have any control over their working environment. This is particularly the case in large, sealed, mechanically ventilated buildings with large areas of open plan. Working environments are usually established before a user's arrival, or a management decision has derived what is an appropriate workspace, with legislation dictating the appropriate temperature, lighting level, desk height, etc. Occupants feel they have no input in determining an appropriate environment for their individuality, group or department to achieve their goals or jobs. It is generally accepted by building occupants that they have no control over their working environments, whether immediate or corporate. If the occupant who is suffering from SBS-related symptoms, does make an attempt to question the working environment set up and this is ignored, or the request is delayed, then symptoms may not only persist but become exaggerated.

This will increasingly bring higher levels of absenteeism which will affect the organization's productivity level, and thus the cost and efficiency of that organization. Cost is really the only method of persuading large corporations to acknowledge that their buildings have a problem and to act on it; the problem is to highlight the direct relation of productivity to pounds and pence. This direct relationship has been established

by Dr Wyon [14] who has determined that the use of individual environmental control systems can increase productivity by up to 7%.

With this in mind not only can productivity be increased by giving
individuals more control over their working environments but also the
amount of absenteeism through building-related symptoms associated
with SBS can be reduced, as building occupants will feel in control of
their working environments and become less susceptible to prolonged
suffering from SBS-related symptoms.

SICK BUILDING OR SICK ORGANIZATION?

As previously discussed, historically when examining the factors behind
mass psychogenic illness or the wider phenomenon of SBS, this has usually been approached as mainly an environmental issue (e.g. relating to
ambient factors such as temperature, lighting, and air flow/quality).
Within this context SBS seems to be treated as a linear phenomenon, as
opposed to a systemic one, where there are (1) a myriad of factors working in an interrelated manner, potentially influencing employee health;
and, more specifically, (2) certain situations creating a certain pattern of
behaviour or response, and investigation focuses on how that pattern is
influenced.

Therefore, when individuals indicate that they are being adversely
'influenced' by the environment within a building, it is important to first
understand the part the individual plays in the wider organizational system, how key elements in the system may be influencing the individual,
and the role the individual may be playing.

In organizational behaviour terms, the organization system comprises
three levels: organization, group, and individual. The factors which comprise each of these levels are living organisms [15] (e.g. people), which
therefore create a larger living system. Just like any other living system, it
is faced with coping with internal/external demands and influences
which can ultimately influence the quality of its own performance. Just as
there are complex balances at work in many chemical reactions and
processes, so are there within organizations, and more succinctly, within
the people that comprise them.

To this end, if individuals claim they are experiencing 'illness' within
the context of their working environment, this could be related to an
actual physical phenomenon in their working environment. That said, it
is also highly probable that the organization itself and wider groups
within it are also experiencing some type of 'illness', and that in reality
the organism as a whole is 'ill'.

THE CONSTRUCTS OF A HEALTHY ORGANIZATION

How can one ascertain the relative 'health' of an organizational system? Broadly speaking the 'health' of an organization could be defined as the ability to cope in the business environment, through the understanding of the processes, as well as the conditions required (physcial and otherwise), to support its ultimate effectiveness/productivity. Those processes and conditions broadly comprise:

- systems structure: formal relationships, management styles, norms and communication (this is defined by most management theorists as the 'culture' of the organization);
- motivation and advancement frameworks: span of control, hierarchies, promotion policies, remuneration/pay systems, etc.;
- information and feedback control: conflict processing, assessment procedures, and the behavioural outcomes from these, and the physical environment.

An organization in optimum health has systems structures, motivation and advancement frameworks, information feedback and control systems, and a physical environment that drive the business to achieve its vision (e.g. where the organization wants to be), mission (e.g. what it wants to do), and strategy (e.g. how the organization tactically wants to achieve its vision). More importantly, individuals will possess a clear understanding and 'buy in' with the company vision and strategy, as well as holding an intrinsic belief that their jobs add value to the business, that they add value to their jobs, and that they have the personal power to act on any of the above (in terms of ability to change/influence events, or to take the active decision to leave the organizational system).

This perspective goes further in supporting the theory that in order to create a healthy organization, attention must be paid and interventions used that equally examine the health of all parts of the organizational system, given they are interrelated and their collective performance is what ultimately ensures effective organizational performance and health.

WORKPLACE-RELATED ILLNESSES AND ORGANIZATIONAL DYNAMICS: SCAPEGOATING

One of the key factors supporting a systemic approach to assessing SBS springs from the behavioural phenomenon of 'scapegoating'. Scapegoating is simply the dynamic whereby an individual (or a group or organization) unconsciously colludes in the process of projecting its anxiety on to a transitory object. This anxiety can be driven by a number of factors, such as loss of control, low esteem or perceived worth, or even more specifically, job dissatisfaction. The transitory object could be something as simple as another individual, or as intangible as a physical environment or building.

Savery [16] undertook an illuminating piece of research which asked the pivotal question 'What does low job satisfaction within the job do to an individual? This made a significant contribution to understanding the phenomenon of workplace-related illnesses in a more systemic framework. The results show that individuals with low job satisfaction feel restless and unable to concentrate, feel irritable, and are depressed and remorseful (see Table 5.1).

Savery goes on to conclude that 'the effect of low job satisfaction on an individual makes the person feel unhealthy … this possibility is further enforced by feelings of depression and remorsefulness about their work which will not encourage him/her to go into work when they feel even slightly ill'. The results of Savery's study show that high job dissatisfaction leads to frustration, and that if the individual concerned feels powerless to leave or change the situation, this scenario can lead to actual ill health. The poor health of the individuals identified as having low job satisfaction was exacerbated by the higher possibility that these people smoked (during working hours, after working hours, and during the weekends – 0.2128, $p = 00.004$; – 0.2139, $p = 0.004$ and – 0.2236, $p = 0.003$;

Table 5.1 Symptoms of individuals with low job satisfaction

Reaction type	r
Feel restless and unable to concentrate	−0.4304**
Feel irritable	−0.4693**
Loss of appetite	−0.0741
Have headache	−0.2250
Feel depressed or remorseful	−0.4047**
Have shortness of breath and sighing	−0.1015
Feel tired, low energy, excessive fatigue	−0.2621**
Feel tense, uptight, fidgety, nervous	−0.2879**
Have difficulty going to sleep or staying asleep	−0.0652
Have upset stomach or intestinal problems	−0.2864**
Have misdirected anger	−0.3002**
Have sweaty and/or trembling hands	−0.2193*
Smoke too much	−0.2487
Have difficulty getting up in the morning	−0.3426**
Feel dizzy or lightheaded	−0.0061
Are you bothered by your heart beating hard?	−0.0608
Do you feel somewhat apart even among friends?	−0.2526**
Do you feel in good spirits?	−0.3762**
Do you feel healthy enough to carry out the things you like to do?	−0.3078**

** significant at $p < 0.001$
* significant at $p < 0.005$

respectively) and consume alcohol more often than their more satisfied colleagues [17].

ORGANIZATIONAL HEALTH AND ORGANIZATIONAL CHANGE

Much of the increased discussion and reporting on the phenomenon of workplace-related illnesses is probably correlated with the increased changes that organizations have been facing over the past eight to ten years in the current business environment. Where organizational change creates widespread ambiguity and uncertainty, resistance to change is likely to be manifest. The resistance is not to change as such – rather it is to the personal loss (of, for example, control, esteem, autonomy, seniority, etc.) that people believe will accompany the change. Therefore change is a dynamic that will influence the equilibrium and homeostasis of the organizational system. This resistance will probably manifest itself – in some cases – as psychological reactance, which will more than likely trigger the increased reporting of workplace-related illness, where the building/physical environment becomes the transitory object.

A number of specific individual factors have now been identified as sources of resistance to change, and which are likely to trigger scenarios of psychological reactance:

- Habit: a change to well-established procedures and practices could create discomfort and resistance on the part of a person who is very familiar with the current system. Inevitably this person is expected to make an extra effort (without necessarily receiving extra remuneration) to learn the new mode of operation. It is therefore understandable that this situation could give rise to resistance to the proposed change.
- Security: doing things in a familiar way brings comfort and security, and people are likely to resist change if they perceive their security to be threatened.
- Economic considerations: People may fear that change could threaten the very existence of their jobs as presently constituted, and eventually lead to the loss of a salary or wage. As a result they resist change.
- Fear of the unknown: Some people fear anything unfamiliar. Any disruption of familiar patterns within the organization, such as changes in reporting relationships, may create fear. This may arise with people thinking that their flow of work will not be as smooth and as fast as previously because they believe it will take time to get to grips with the changed arrangements.
- Lack of awareness: due to selectivity in perception, a person may overlook a critical facet in a change process. It could be that the facet ignored – e.g. requiring a double signature on travel expense claims –

is something the person is opposed to, and somehow it is conveniently overlooked. As a result, there is no change in the person's behaviour (at least initially) as far as the changed practice is concerned.
• Social considerations: the motivation to resist change may spring from a group. If a change to rules and regulations was unilaterally imposed by management, but resisted by a work group, a member of that group may oppose the change simply because acceptance of the change could amount to the disapproval and perhaps be subjected to the application of sanctions operated by the group [18].

LOCUS OF CONTROL AND RESISTANCE

As discussed earlier, the perception individuals hold with regard to how an environment can affect them is driven by their personal locus of control. In regard to locus of control and organization change, a prevailing view is that people who define events in their lives as being outside their control (i.e. external locus of control) will be less able to cope effectively with stress (especially that brought about by widescale organizational change or transformation), and are therefore more likely to experience physical distress than people with internal locus of control beliefs [18].

A MULTIDISCIPLINARY APPROACH TO THE ASSESSMENT OF WORKPLACE-RELATED ILLNESS

With the rise of interest and the success associated with a more multidisciplinary approach to improving organizational performance and to organizational problem-solving, organizational development theories and interventions provide a credible blueprint on which to build a workplace related illnesses research methodology. Organizational development is defined as 'the process of planned change and improvement of organisations through the application of knowledge of the behavioural sciences. More specifically, organizational development can be described as 'an involved network of events that increases the ability of ... the organization ... to solve problems in a creative fashion, to assist ... in adapting to the external environment, and to manage organizational culture. It embraces a broad range of interventionist processes from changes in organizational structure and systems to psychotherapeutic (oriented) counselling sessions with individuals and groups in response to changes in the external environment [18]. These interventions seek to improve the effectiveness of the organization and contribute to employee well-being.

Bringing this down to a more tactical level, it seems fair to surmise that a holistic approach to examining SBS could increase the probability of identifying the root causes of SBS as opposed to just focusing on the symptoms of the problem. This holistic approach would be based on the recognition that there are three broad factors within the context of an organizational system that must be examined when looking to assess the nature of SBS:

- environmental factors, such as comfort/layout of work spaces and work areas, as well as ambient factors such as temperature, lighting, air flow/quality, and IAQ/HVAC systems;
- psychosocial factors: dealing with work/organization/flow, nature of work, stress/performance factors (including management styles, culture, and organizational norms); and
- physiological factors: the impact of the above on individual's (or groups') physical and behavioural attributes.

The credibility of taking this type of approach to identifying and ultimately eliminating workplace-related illnesses is supported by the fact that in an analysis carried out by Jerry Porras (Stanford University) and Peter Robinson (University of Southern California) of a range of empirical organizational development research projects undertaken between 1973 and 1988, projects centring on the changes to the physical setting resulted in very little positive change in any of the variables (e.g. organizing arrangements, social factors, technology, physical setting, individual behaviour, organizational outcomes), except for the physical setting variables themselves [19].

Porras and Robertson [19] conclude it could be the lack of use of physical setting interventions in organizational development that has prevented the accrual of much information regarding the most appropriate changes to make or the most appropriate method of implementing them, and that as a result these interventions are less effective than other types. However, in relation to the phenomenon of workplace-related illness, Porras and Robertson comment 'the fact is that physical setting interventions result in negative changes more frequently than other types of interventions' (p. 789). Most importantly, they go on to state this is likely to be the case because of the unpredictability of human behaviour, and that behaviour is influenced by numerous organizational factors. Changes or interventions focused on one set of factors cannot be guaranteed to produce the desired consequences, and in fact may result in the opposite. They therefore conclude that

it is important to plan and implement multiple types of interventions, such that organisational characteristics will consistently affect behaviour in desired directions. In regard to which types of interventions created the highest rate of positive change in organisational outcomes, it was organising arrangements (structurally and work/task related) and social factor interventions (e.g. improvement of interaction processes—such as 'team building', better understanding of informal power structure, etc.) that proved the most effective. These interventions both produced positive change over half the time, and neither of them resulted in any negative change ... the negative impact of physical setting interventions on individual outcomes was the highest of any intervention on any category of dependent variables (p. 790).

CONCLUSION

Given that an organization is a complex and interrelated living system, the phenomenon of workplace-related illness is not likely to be eradicated by looking at only the environmental and physical aspects of the building which the organization occupies. In order to truly start to understand and create viable solutions to SBS, organizations need to take a systemic approach to assessment, which involves adopting a strategy and methodology for investigating psychosocial phenomena which may also be contributing to the reports of SBS symptoms at all levels in the organisational system. Within this it is also critical for organizations to establish a working understanding of the main types of workplace illness, to reduce the risk of misdiagnosis and inappropriate treatment interventions.

Organizations that choose not to investigate the behavioural and psychological factors, and rationalize the issue as a simply environmental one, will more than likely be faced with spending significant monies on treating symptoms, as opposed to root causes, which inevitably will lead to the continuing plague of reported SBS incidences.

REFERENCES

1. Ryan, C.M. and Morrow, L.A. (1992) Dysfunctional buildings or dysfunctional people: an examination of the sick building syndrome and allied disorders. *Journal of Consulting and Clinical Psychology*, **60**(2), 220–4.
2. Hartman, D.E. (1988) *Neuropsychological Toxicology*. Pergamon Press, Elmsford, NY.
3. Hodgson, M.J. (1989) Clinical diagnosis and management of building-related illness and sick building syndrome. *Occupational Medicine: State of the Art Reviews*, **4**, 593–606.
4. Kreiss, K. (1990) The sick building syndrome: where is the epidemiological basis? *American Journal of Public Health*, **80**, 1172–3.
5. Colligan, M.J. and Murphy, L.R. (1982) A review of mass psychogenic illness in work settings, in *Mass Psychogenic Illness: A Social Psychological Analysis* (eds Colligan, Pennebaker and Murphy), Erlbaum Hillside, NJ.

6. Hall, E.M. and Johnson, J.V. (1989) A case study of stress and mass psychogenic illness in industrial workers. *Journal of Occupational Medicine*, **31**, 243–50.
7. Molhave, L. (1990) The sick building syndrome (SBS) caused by exposure to volatile organic compounds (VOCs), in *The Approach to Indoor Air Quality Investigations* (eds Weekes and Gammage), American Industrial Hygiene Association, Akron, OH, pp. 1–18.
8. Whorton, M.D., Larson, S.R., Gordon, N.J. and Morgan, R. (1987) Investigation and work-up of tight building syndrome. *Journal of Occupational Medicine*, **29**, 142–7.
9. Burge, S., Hedge, A., Wilson, S., Bass, J.H. and Robertson, A. (1987) Sick building syndrome: a study of 4,373 office workers. *Annuals of Occupational Hygiene*, **31**, 493–504.
10. Robertson, A.S., Burge, P.S., Hedge, A., Sims, J., Gill, F.S., Finnedan, M., Pickering, C.A. and Dalton, G. (1985) Comparison of health problems related to work and environmental measures in two office buildings with different ventilation systems. *British Medical Journal*, **291**, 373–6.
11. Rubin, Z. and McNeil, E.B. (1983) *The Psychology of Being Human*, Harper and Row, London.
12. Brehm, J.W. (1956) Post-decision changes in the desirability of alternatives. *Journal of Abnormal and Social Psychology*, **52**, 384–9.
13. Seligman, M. (1972) *Biological Boundries of Learning*, Appleton-Century-Crofts, New York.
14. Wyon, D. (1995) *Individual Microclimate Control: Required Range, Probable Benefits and Current Feasibility*, White Paper.
15. de Board, Robert (1978) *The Psychoanalysis of Organisations: A Psychoanalytic Approach to Behaviour in Groups and Organisations*, Tavistock Publications Limited, London.
16. Savery, Lawson K. (1989) The influence of job factors on employee satisfaction'. *Journal of Managerial Psychology*, **4**(1).
17. Singe, Peter M. (1990) The Fifth Discipline: The Art and Practice of the Learning Organisation, Doubleday, New York.
18. McKenna, Eugene F. (1994) *Business Psychology and Organisational Behaviour: Student Handbook*, Lawrence Erlbaum, Hove, East Sussex.
19. Porras, J. and Robertson, P. Organisational Development: Theory, Practice, and Research, Gower, Aldershot, Hants.

CHAPTER 6

Maintenance

Robert Davies

INTRODUCTION

This chapter is aimed at the building or facility manager who has limited experience of building services. The building or facility manager performs many functions including personnel management, financial control, etc. with the maintenance function usually an 'add-on' responsibility. The manager is usually from a non-engineering background so he needs to be aware of such matters. The chapter will cover routine maintenance aspects, the neglect of which will undoubtedly cause environmental problems.

SBS has not been tied to one specific area but research [1] has shown that many different factors cause many people problems. If the maintenance of a building is carried out effectively some of these problems will be eradicated. However, good maintenance does not make a badly designed system better. A heating, ventilating or lighting system, that by faulty design, does not produce conditions in the 'comfort criteria' for a space, will not be improved by maintenance but only by a redesign and refit. The main areas that have been identified as possible causes of SBS are:

- lighting;
- heating, ventilating and air conditioning;
- processes carried out in the building, which includes chemicals, etc.;
- actual working and environmental conditions.

As can be seen from the list, there are many possible areas of concern. Eradicating one problem may help several people but will not guarantee 100% comfort for all occupants.

MANAGEMENT

The effective management, in terms of engineering, of a building is very important. Should the management hierarchies and procedures implemented be unsatisfactory, there is the possibility of problems with the day-to-day running of the building services. A clearly laid down management structure must ensure that the plant is properly run and maintained to keep the building in comfort conditions for the occupants. This aspect of the running of a building is often neglected for several reasons – mainly cost – and the result is an inefficient and ineffective heating and ventilating system. Should an owner look back to the days when he procured the building, he will find that the lion's share of the cost was borne by the building's services. To ignore this is to ask for problems after the halcyon days of guarantees and warranties.

There are usually only three ways a building's services are maintained:

1. the employment of maintenance contractors;
2. direct labour from within the organization;
3. a hybrid of both methods.

All these methods have their own advantages and disadvantages that will be discussed more fully.

Maintenance contractors

This is now universally seen as the best method of providing maintenance to a building's services. The advantages are quite obvious; personnel are only used as and when required. In this method of maintenance, the management and supervision is carried out by an engineering contractor who supplies everything to do with the task. This usually involves having a fitter, or operator, heavily involved with the day-to-day running of the plant. He carries out minor repairs but has on call all the necessary back-up trades should a problem occur that he cannot deal with. One of the main 'breakdown' problems that occur with plant is the tripping of fuses and relays. An operator can easily reset these; but has he cured the problem that caused the relay to trip? True, he has kept the plant running, but for how long? Will the next breakdown be a large one because the first one was not repaired properly?

On the debit side, a building's engineering plant and services is unique. There are very few 'standard' plants, so an operator needs considerable experience of working on a system to get to know it and how it works; with contract maintenance, this does not tend to happen.

In-house maintenance

This was the traditional method of maintenance until the 1980s, when the emphasis changed to a more budget conscious approach. Here the management, supervision, operation and maintenance of a plant were provided by the organization itself. This meant high overhead costs for employed personnel as well as high cost for the storage of spares, etc. This system keeps a lot of capital tied up.

The advantage is that the plant is run by experienced personnel. Both supervisors and tradesmen are aware of potential problems that may arise. It is also easier to get a job done right away should a problem occur. Experience will tell if a certain job needs doing more or less frequently than recommended by the manuals.

Hybrid method

This is a combination of both methods and usually consists of a plant manager, it is hoped an engineer, and several fitters or operators who run day-to-day operations. Any maintenance of a major type required during or out of normal plant working hours is supplied by a contractor. Again, the problem is the contractor's lack of intimate knowledge of the plant.

Whatever the system chosen for maintenance, it is vital that the following principles are followed:

- planned maintenance schedule;
- record kept of all breakdowns, repairs and maintenance;
- only recommended spares, oils, lubricants and filters are used in the plant;
- correct supervision is carried out on all tasks;
- correct maintenance procedures are carried out;
- equipment is replaced at recommended intervals;
- management checking or monitoring of standards of work.

Too often building maintenance and services maintenance is given as a function to someone without the necessary mechanical/electrical background. It is vital that the manager understands these two areas. However, an in-house electrical and mechanical supervisor can provide valuable guidance.

PLANT MAINTENANCE

The building can be likened to the human body, the building frame being the skeleton, the cladding and roof being the flesh. Then there are the vital organs, the heating and ventilating or air conditioning plant, the electrical installation and the nerve centre or brain – the building management system. Just like the human body, all parts need feeding and maintenance to ensure correct operation. Should any part malfunction, this will affect other parts.

SBS has not been found to be caused by one specific item in a building, but research has found many possible contributory sources [1], all of which involve the use and running of a building with an 'artificial' environment. There needs to be a global building approach to cleaning and plant maintenance to ensure the best possible conditions for the occupants. The outside atmospheric conditions such as the thermal gain, wind factors, including prevailing winds and discharges from other buildings or factories also need to be addressed. These can all give problems to the user of a building. The usage of the building and the pollutants by canteens, dusty files, photocopiers, printers, etc. should also be considered. The general cleaning of the building needs to eradicate pollutants as well as to keep its appearance acceptable. All such issues need to be taken on board and a global approach adopted for the building.

Air filters

There are many types of air filters which must be maintained to the manufacturer's specification at the recommended time intervals [2]. The filters must be able to filter out all pollutants present. So the plant inlet filters have to be fitted to take into account the pollutants in the intake area, which may include not only dusts, but also chemicals. A good proportion of air is recirculated to aid energy conservation; this is acceptable as long as the recirculated air is filtered to remove pollutants. Much dust and many chemical pollutants will be picked up by the air on its circulating journey, and these must be removed. Carbon dioxide levels must not be allowed to become unacceptable. This can be eradicated by introducing fresh outside air. Filters must be fitted correctly so that no air can bypass the filter, and it is vital that the specified filter only is used. At various times of the year, i.e. during dry hot summers, filters may need to be changed more frequently. Filters are the vital first line in the air supply [2].

There are three types of air filter:

1. Dry filters: these are usually disposable and are constructed of compressed paper or fabrics. These need to be changed at recommended intervals.
2. Wet filters: these have a coating on a metal or plastic grille with the dust collected on to it. There are two types, either a roller or fixed panel. Again they must be changed or cleaned as specified.
3. Electrostatic filters: these are the most effective as they remove both small and large particles. Air is passed through an ionizer where it receives an electrical charge. The particles then pass a plate where a charge is held. The particles then stick to the plate. Electrostatic filters also remove bacteria. These filters must be switched off when the plant is shut down. This can be arranged for by inclusion in the plant wiring start-up.

A badly maintained filter can actually pollute an airstream, and can drastically reduce the flow of air through the system. It is essential that the specified amount of air per hour passes through the conditioned space [3].

Humidification

This is an area where problems undoubtedly have arisen in the past. There are two types of humidification: (1) air washers and (2) steam.

Air washers

Air washers are used to either cool or humidify air. This particular method has been found to cause problems due to several factors:

- Water is carried away from the washer area, causing biological growth in the system.
- The excess water pan contains a 'biological tank of growth'.
- Chemicals designed to stop the biological growth enter the air flow.
- Excess water carried away rusts other parts of the plant.

This system is unsatisfactory and should be replaced as soon as possible. However, if in operation, the following maintenance points need to be addressed:

- Use only untreated water in the sprays.
- Sludge and waste in the reservoir need to be cleaned regularly.
- Correct orientation of the eliminator plates to stop water carry over into the system and aid air flow.
- Drip tray to take away eliminated waste to be cleaned regularly to ensure water is not held in tray.
- Spray nozzles to be regularly cleaned to ensure atomization of water spray.
- Ensure there is an air break on the drip tray to stop return siphonage.

Steam humidification

This is now the preferred method of humidification. Here steam is injected into the air flow and taken up. This system gives few problems and there is usually no carry over of water into the system. It is important that steam contains no additives, such as those used in steam boilers to prevent corrosion. There are steam humidifiers on the market which use water direct from the mains. Points with regard to maintenance are:

- Atomizer spray mechanism is regularly cleaned to remove any build up of dust, etc.
- The control mechanism enables just the correct amount of steam into the air flow, eliminating drips on the nozzle.
- Remove scale build up in hard water areas.

Cooling plant

The cooling plant consists of three main parts:

- cooling coil;
- pipework to coil and cooling mechanism;
- cooling mechanism either air-cooled condenser or cooling tower.

The cooling coil in the plant cools the air flow as it passes over. This causes condensation on the coil as water leaves the air stream. The water runs down the coil on to a drip tray. It is important that no water is held in the tray, or this could be carried back into the air flow or aid bacterial growth. Points for maintenance are:

- Drip tray is clean and falls to drain are clear.
- Drain is unblocked and has an air break to stop siphonage and bacterial growth.
- All fittings carrying water are watertight and no plant water is carried into the air stream.
- All deposit build-up from previously leaking joints is removed.
- Cooling towers are now being replaced extensively by air-cooled condensers. These condensers must be maintained effectively by contractors or in-house personnel. To reduce the efficiency of the refrigeration plant by a small amount, means the massive loss of cooling capacity causing the building to possibly overheat. Should there be problems keeping a building in 'comfort criteria' with regard to temperature, a calculation needs to be carried out to see if: (1) there is sufficient cooling plant available in the present system; and (2) to identify and classify all additional cooling loads since the cooling system was designed.

Heating

This is carried out by a coil placed in the air stream within the plant. The air is heated as it passes over the coil. The points to note are:

- The coil is not leaking at joints causing plant water to enter the air stream.
- The control valves on the water flow are calibrated and controlled correctly.
- The valve is operating correctly with no leaks on the gland.
- The control valve is receiving correct information to act on. This requires a check in the building management system that sensors, thermostats, etc. are operating correctly.

Should a control valve be stuck open or closed, the results are obvious. The heating water needs to flow into the coil at specified temperature and flow rate. This can sometimes be a problem on very cold days and this is why an 'overcapacity' is designed into the system. Should this be a problem, checks need to be carried out for additional heating loads added since the building was designed. There are sometimes additional loads placed on the heating system where, for example, the conditioning units at the occupied areas can have an additional heat load put in. This is called terminal reheat. Here the points to note for maintenance are:

- Are the flow and return valves working correctly?
- Is the thermostat controlling the unit placed correctly, calibrated correctly – or even there?
- Is the unit itself serviceable?

Should there be units in the space that provide heating individually, they must be maintained to retain efficiency. The three main areas are:

- control mechanism;
- heating/cooling mechanism and pipework;
- actual distribution method, e.g. fan.

This will require maintenance in both electrical and mechanical areas. Maintenance must take place to manufacturer's specification and time periods in the occupied space. Should the thermostats be placed incorrectly on a cold spot on a wall, too high, in an area where there is high thermal gain from electrical equipment or solar activity, in a storeroom, covered by files or books, damaged or neglected, or in an area not representative of the human occupation regime, they will provide incorrect information to the plant and that room will be serviced incorrectly. These problems need to be addressed and maybe the sensor moved. As a part of normal maintenance, the thermostat must be maintained and calibrated.

Humidistats

The criteria for thermostats are the same for humidistats. The readings for these instruments may be checked by hand measurements using a sling sychrometer. This is an easy and convenient method to use.

Air circulation plant

This consists of mainly ductwork, both fixed and flexible, plus a fan or fans. This part of the system is often neglected as it is never seen.

Fan

This actually pushes air around the plant and ductwork, though a simple piece of equipment incorrectly installed can cause problems. With regard to maintenance, the main points are:

- lubrication of bearings;
- balance of the fan.

The fan may be sealed, in which case it will require no additional lubrication. The efficiency of the fan depends upon its ability to move without excessive friction. This is enabled by lubrication. When lubricated, all excessive grease must be removed to stop grease entering the air system. Should a fan be unbalanced, excessive vibration will be noticed.

The fan is usually turned by an electric motor and belts or a drive mechanism. The belts must be in good condition and correctly adjusted to ensure the fan is turning at the correct revolutions per minute. Should a drive mechanism be used, this may require maintenance as per the manufacturer's specification. Excessive build up of dirt, grease, etc. must be removed from the fan and blades as this will affect the balance of the mechanism. The caging should be inspected to ensure the blades are not rubbing.

There are fans for both inlet and outlet air, i.e. inlet and exhaust fans. Both need maintenance; arguably the exhaust fans need more, as they operate in dirtier conditions. Fans operating at fewer revolutions per minute than specified mean less air entering the building.

Ductwork

This needs inspecting periodically to ensure the integrity of the system. There is little chance of contamination entering the ductwork as it is under positive pressure, i.e. higher pressure than in the rooms. However, all joints must be airtight to prevent expensive conditioned air being lost into roof voids, plant rooms and ceilings. Joints between fixed formed ductwork and flexible couplings to air diffusers need to be checked for airtightness. All holes, rips and tears in flexible couplings must be sealed. All pressure, air temperature or other sensors that are fitted into the ductwork must also be sealed. Should any corrosion be noticed on ductwork, it must be investigated to ensure that there are no leaks:

- from plant in the system (heating humidification or cooling);
- from the building's outer skin;
- from hot or cold water services to toilets.

There are inspection panels on the ductwork to allow access to their interior. These must be checked and the panels replaced and resealed. Should there be excessive dust noticed in the ductwork it indicates a problem with filtration. All dust should, if possible, be removed, along with any other foreign bodies like cups or newspapers! The mixing plenum for fresh and recirculated air needs regular inspection and cleaning. The internal state of the ductwork is a good indication of the care of the plant. All loose nuts and bolts need tightening as they will eventually wear through the metal.

Air inlet/extract diffusers

These need regular maintenance to remove dirt and any other obstructions. They should not be restricted in any way by either furniture, paper, etc. On no account should they be covered, as this will reduce air flow. All ventilated and conditioned spaces need at least one inlet and one outlet diffuser. Should this not be the case, check that partitioning has not been erected, breaking up a space. Partitioning should not be placed over diffusers. When replacing diffusers check that the air is being circulated in the correct manner about the room. The positioning of the grille is important.

Control equipment

This is an area very often undermaintained or even ignored. The positioning of sensors is very often poor; this will provide the plant with inaccurate information, leading it to supply the conditioned space with incorrect air.

Thermostats

Many types of thermostats are in use throughout a building. The thermostats in the plant, e.g. boiler, can be checked and calibrated easily by the contractor's engineers and this should obviously be done as per the plant manufacturer's schedule. However, the main problem is usually remote thermostats in the area where occupants are active.

General plant controls

These will be covered by the manufacturer's schedules and must be maintained effectively. Wrong feedback on what is happening in the plant will affect comfort conditions. Once calibrated correctly, the operation of automatic controls, like automatic gate valves, needs checking to ensure correct operation. Should a building management system be in operation, this makes the task easier, as plant feedback in terms of temperatures, humidity, etc. is readily available and displayed quickly. The main areas for concern are:

- calibration of instruments;
- operation of mechanical control mechanisms, often easily done by lubrication.

All feedback information must be available to the system. It is of no use if a vital measurement is missing from the plant control board as the system will make a wrong decision. For this reason, all sensors must work, be calibrated, and be operating during plant running times.

LIGHTING MAINTENANCE

Buildings are lit by electricity and as part of Building Regulations must have a natural lighting component. The building's lighting system [3] is a complex piece of electrical engineering; but once in place, the maintenance is usually straightforward, just needing planned replacement of parts and cleaning. The lighting system will be a mix of general space lighting and specific task lighting. There may be a light meter control which switches off the general space lighting at various times when the natural light brings the space up to design criteria lighting levels. These areas need assessment and a suitable maintenance schedule introduced.

General space lighting

This is now almost universally by fluorescent tubes in diffusers which provide an overall lighting level in a space. There are three main parts of the system:

- the actual lighting tube;
- the control mechanism;
- the diffuser.

The lighting tube gives off light which is directed into the space by the diffuser. There are many types of fluorescent tubes which operate throughout the visible light spectrum. The visible light spectrum is from infrared, which is associated with 'warmness' to the ultraviolet end which is associated with 'coolness'. The choice of tube is important; care must be taken that tubes which are replaced are to the original design concept. Very often the 'feel' of a room is a result of the lighting used.

A fluorescent tube has a lighting life of so many hours, which is specified by the manufacturer. It is vital that the tube is replaced at this point or the lighting efficiency of the tube drops. The lighting life of the tube is also affected by the number of times the tube is switched on and off. Should an automatic control system be used, this will reduce the planned life of the tube. The tube needs to be dusted regularly to produce maximum output.

Diffuser

This actually holds the tube – the control mechanism – and distributes the light. Diffusers need no maintenance apart from regular cleaning, to ensure that the light is distributed evenly and effectively. Very often the diffuser acts as a return vent for exhaust air. This has several advantages including cooling the tube and removing heat from the space.

Control mechanism

To start and run the tube, an electric starter and choke are fitted for a current boost when the light is switched on. These must be replaced at recommended intervals to ensure efficient running of lights.

When equipment is worn, humming and flicker of lighting tubes enters the space and this is a cause of SBS. Good maintenance is essential.

Individual task lighting

This can be used in individual work spaces so that the occupant has control over his or her own working environment. This should be encouraged; but care needs to be taken that the lighting level is correct for the task. The lighting level needed for VDU operation is very different from that at a draughtsman's board.

VDUs are becoming increasingly common in all working spaces, and it has been found that a reduced overall lighting level in the space reduces glare on the screen and that localized task lighting aids comfort conditions. Where there are VDUs, special diffusers are available to reduce or even eliminate glare. Uplighters also provide diffuse light with no glare, using the ceiling as a diffuse light source.

General electric services

General lighting levels in a space are important to aid comfort conditions. In older buildings, the electrical system may not be large enough to cope with new electrical office equipment such as VDUs, printers, fax machines, etc. This may cause a voltage drop on the electric lighting circuits which will reduce the efficiency of the tubes, causing lighting levels to fall. Ideally, the lighting circuit on a floor level should use all three electric phases, as this reduces the problems of flicker. This also helps to spread the electric load more evenly. All electric sockets in the work space should be operative so that individual sockets are not overloaded causing voltage drops in small areas.

CLEANING

Cleaning of a building is a necessary function; it keeps the building pleasant to work in, free from dirt, odours and removes some contamination sources.

Photocopiers

Photocopiers produce chemical pollutants which are often introduced into the working environment. They need to:

- be isolated from general office areas in a sealed room;
- have an independent means of exhaust of pollutants, separate from the building exhaust to prevent recirculation;
- be cleaned regularly to remove chemicals and paper dust.

Printers

These are essentially the same as photocopiers; however, if they are in a work space they need to be in a sealed box to provide acoustic insulation. They need to be regularly cleaned to remove paper dust.

Hard working surfaces

These need daily damp dusting. Dry dusting is inappropriate as dust is disturbed and resettles again. All files need to be in enclosed cabinets as they are notorious dust holders. Filing cabinets should be regularly vacuumed.

Electrical equipment

VDUs, computer equipment, phones, etc. are notorious dust holding areas. They need to be damp dusted daily to remove all deposits. The removal of dust from VDUs improves the vision quality of the screen. Glare screens need vacuuming carefully.

Soft furnishings and carpets

Chairs, carpets and blinds require vacuuming on a daily basis. Curtains need regular washing to remove dust. A powerful cleaner should be used to ensure all dust is removed.

Hard floor surfaces

These need to be brushed, dampmopped and polished. Dust settles on these surfaces. Just brushing puts the dust back into the atmosphere. Damp mopping removes this dust. Polishing helps to remove dust; however, buffer pads hold an enormous amount of dust and require regular washing and replacement. In a working environment the following areas need regular dust and dirt removal by dampdusting methods and vacuum cleaning:

- inlet and exhaust louvres;
- light fittings;
- filing cabinets;
- carpets and soft furnishings;

- hard working surfaces;
- hard floors;
- electrical equipment.

Windows

Windows need to be cleaned internally and externally for both aesthetic and comfort criteria reasons. They should be cleaned at least twice a year, more preferably. Dirty windows stop natural light entering a space, thus reducing natural lighting levels in the space. Windows should be draught free and sealed to stop dust entering the space.

External blinds

If fitted these must be operable as they are a good method of stopping internal glare, as well as in summer conditions cutting down on solar heat gain into a space, thus reducing the cooling load on a building. Internal blinds help but are not as effective as external blinds.

Decoration

Newly painted and decorated surfaces make a room pleasant to work in and increase the lighting levels. Clean ceilings are important when uplighters are used as the ceiling surface is used as a diffuse light source.

Maintenance of cleaning equipment

As with any mechanical equipment, cleaning equipment should be maintained correctly – it is usually not maintained at all. A vacuum cleaner that is not working as it should is a potential hazard, as the dust removed is blown straight back into the space.

The main areas of concern are:

- Dust bag: is it sealed properly to the extraction tube? Is all the dust contained in the bag and not allowed to escape? For this reason only recommended dust bags should be used. They should be sealed correctly and changed regularly.
- Brushes: are brushes that are in contact with the carpet, worn? Should this be the case, then the cleaner is not working efficiently.
- Quality cleaner: the cleaner should be of a commercial quality with a powerful robust motor.
- Can it clean all required areas? Can the cleaner get under desks and into corners or will special tools be needed? Floor polishers are generally quite easily maintained. The only piece of equipment that needs inspection is the buffer ring. This must be cleaned regularly to remove

all dust and replaced when worn. Dusters need washing or replacing regularly. Bad cleaning is the same as no cleaning.

REFERENCES

1. Raw, G. (1992) *Sick Building Syndrome: A Review of the Evidence on Causes and Solutions*, HSE Contract Research Report No. 42, Building Research Establishment/HMSO, Watford.
2. Jones, W.P. (1988) *Air Conditioning Engineering*, Arnold, London.
3. CIBSE (1986) *Guides Vol. A & B: Design Data and Installation and Equipment Data*, Chartered Institution of Building Services Engineers, London.

CHAPTER 7

Human resource management

Chris Baldry, Peter Bain and Philip Taylor

INTRODUCTION

The office: building and institution

'The office' is one of those terms, like 'bank' or 'school', which refers both to a particular type of social and economic organization and also to a recognizably distinct type of building [1]. The structure of the office building and the organization of office work have maintained a close relationship since the eighteenth century when early office functions were dispersed throughout the city and offices were little more than rooms in merchants' houses or coffee houses [2]. As the idea of a fixed office location emerged in the nineteenth century, with the growth of such organizations as banks and insurance companies, at first they still resembled private houses or developed into the 'chambers' model of small individual offices in a single building [3]. In this nineteenth-century counting house the predominantly male 'black coated' workers enjoyed the gentlemanly status (if not always the income to match) afforded by the scarce skills of above average literacy and numeracy [4].

New forms of organization at the end of the nineteenth century, such as mail order business, the railways and new utilities (gas, electricity), followed by an expanding public sector (post offices, pensions, health insurance) brought with them a much larger scale of office organization in which the principles of F.W. Taylor's scientific management were rapidly applied. Work in these offices was hierarchically ordered, routinized and subject to a minute division of labour; office workers were losing their former status and office work was becoming an increasingly feminized sector of the labour market. These mass offices, and the development of the 'modern' open plan office in the mid-twentieth century, were made possible by continuing technical developments in steel frame construction, and later heating and ventilation, which allowed for the open, deep space necessary for the visual superintendence of the new office workers [3], while the work itself was facilitated by developments in the technology of information handling such as the typewriter and telephone.

After 1945 the number of white-collar jobs increased in tandem with a massive post-war expansion in public services provision and the advent of much larger companies, so that by the 1970s, white-collar workers accounted for around half of all those in employment in the major economies. These changes to the composition of the workforce were reflected in the construction of huge new office buildings and the property boom of the late 1950s and early 1960s [5], resulting in the construction of low quality speculative buildings for sale or rent, with an emphasis on exteriors which would appeal to corporate customers as symbols of power and prestige.

However, these developments were contributing to a double crisis. First, the burgeoning labour costs of undercapitalized offices had created a bottleneck in productivity compared to the gains which early 'hard automation' had made in manufacturing production. Secondly, the oil crises of the 1970s led to a series of substantial hikes in the cost of space heating and lighting: office buildings, from being seen primarily as 'prestige' showcases of corporate power had suddenly become expensive. The crisis was temporarily resolved in two ways. The development of information technology towards the end of the decade fortuitously held out the promise of productivity gains to match those on the shop floor but, prior to this, substantial changes had taken place in the methods of office construction. The new energy cost awareness in the climates of Northern Europe and North America dictated an emphasis on increased insulation and the construction of 'tight' buildings with no openable windows, a heavy reliance on recirculated air for both heating and ventilation, centrally controlled by building management systems which were themselves increasingly computerized. These shifts in the priorities behind office construction took place at a time when very significant changes in the nature of office work were under way.

The restructuring of office work

Over the past two decades the nature of contemporary office work has clearly undergone some profound changes. In a climate of sluggish growth and increased international competition in the private sector and an ever-tightening squeeze on finances in the public sector, organizations have become almost desperate in their search for competitive advantage and economic survival. Some have seen the solution to be the delivery of quality services and products through total quality management, others through the adoption of human resource management strategies. Other organizations have used a period of sustained high levels of unemployment to derecognize trade unions, dispense with collective bargaining and adopt new 'flexible' labour market policies with a greatly increased use of temporary and part-time employment.

However, the phenomenon which now dominates every aspect of the organization of office work is undoubtedly information technology (IT). The function of the office has been described as being a combination of communication and the control of complexity [2], functions which of course make office work pre-eminently suitably for computerization. Castells [6] suggests that this has affected the office in several ways.

First, IT contributes to the 'flatter' organization, with fewer layers of hierarchy as layers of middle management are no longer required. The move away from bureaucratic control mechanisms means that there is less requirement for the hierarchical ordering of space. Open plan offices were always designed to ease the task of supervision [1, 3] but initially only the routine clerical functions were open plan, while management and supervisors retained their separate spaces. Duffy [7] argues that, as organizations restructure to manage with fewer hierarchies and to work with groups or teams, so space is allocated more according to the criterion of functional need rather than by position in the hierarchy, there will be more acceptance of shared space, of open-plan and undifferentiated workstations so that increasingly offices are being designed as a flexible shell. For example, in a study in the early 1980s Baran noted how restructuring in the insurance industry associated with computerization had resulted in the tearing down of private office walls, and how the new teams of professionals, managers and clericals now worked side by side separated only by movable partitions [8].

IT also makes possible the functional and spatial division between HQ, back office and customer 'face-to-face' offices; these can now be geographically separate, with office locations chosen for specific features, such as type of labour supply [9], appropriate to each function, but all linked by the electronic information flow [10]. In addition, there has been much discussion of the possibilities of home working or remote working affecting an increasing proportion of white-collar workers, leading some to predict the coming of the 'dispersed organization' and the 'virtual office' [11, 12].

The aspect which has probably received the most attention is the use of IT in the organization of work itself. There is no doubt that this has been a mixed blessing: the new technology has led to more varied and challenging jobs as IT makes possible the reintegration of formerly functionally separate tasks, now reunited under team working [6]. However, it has also often meant work intensification for the office worker [13] with increased monitoring and surveillance of work performance [14] and, despite a decade of health and safety publicity and campaigning, working with office IT still carries its attendant hazards of repetitive strain injury, eye strain and general stress [15].

New forms of organization and new economic priorities have influenced the emergence of new practices regarding the terms and conditions of employment. The economic success of the Japanese seemed to many to be strongly linked to the commitment and loyalty which the Japanese company apparently expects of its workers, which contrasts sharply with the 'low trust' adversarial employment relations typical of UK and US companies. This and the exhortations of management gurus that the secret of competitive edge lies with the management of people [16] has created a head of steam behind the new language of human resource management (HRM).

There is now agreement [17] that there are two versions of HRM, 'hard' and 'soft' or, more accurately, two different emphases in the mix of practices that are now encountered in organizations. Hard HRM emphasizes managing the head count in the most rational way in order to minimize cost, while soft HRM emphasizes the resource base available to an organization that is to be found in the people who work for it. This highly adaptable resource can be liberated by training and development and by cultivating a high level of sustained employee commitment to the organization [18], often by the promotion of a distinct company culture, stressing the identification of the individual worker with the goals and ethos of the company. There has been a corresponding increase in use of terms like 'empowerment' [19] to describe the devolving of responsibility to the level of the employee and an increased use of the work team as the basic unit of organization. All versions of HRM claim that because of the intrinsic importance of people management, human resources will take on a much more strategic and high profile role in the organization.

Does the building support the work?

This restructuring of office work and the rapid diffusion of information technology have combined to generate at times an almost utopian vision of the 'post-Fordist' office of tomorrow [7] which will be flexible, flat, creative, dispersed and part of a wider information society: a paperless, networked, information exchange housed within an 'intelligent building' [20], in which technology enhances not simply the company's work but also the environment in which it is performed, and whose external appearance conveys a corporate image of success, modernity and progress. Whether or not this vision of the future is fulfilled, there is no doubt that the increasing emphasis on employee commitment will require the built environment to be more supportive of the HR strategies adopted by the organization.

However, when we look at how working in such offices is actually experienced by their occupants things often seem very different; it must be questioned whether the contemporary office building is supporting these organizational developments at all. The phenomenon of SBS is a direct challenge to the idea that working in a modern prestigious building will be a source of pride and motivation to employees. This book and previous investigations of SBS [21–25] are testament to an increasing awareness that the built working environment, rather than enhancing employee commitment, frequently seems to malfunction in such a way as to make its occupants feel constantly unwell. This contradiction between the image of the modern office and the experienced reality, is exemplified in the following two statements which refer to exactly the same office building in Glasgow:

> The building is superbly finished inside and out, providing a high quality working environment for [the company's] staff. Entry to the building is though a dramatic atrium. Office space is on four floors which offer large floorplates where staff can be deployed flexibly … With a massive amount of glass facing southwards, a sophisticated environmental control system ensures that the offices and technology are kept at an ideal working temperature. Equally important, the high-tech wiring under the floors provides [the company] with flexibility to change its computer and communications systems as required.
>
> (Developers' advertising feature, *Glasgow Herald*)

> This is not a sick building, it's a terminally ill building … You can never tell in advance what this building is going to be like. One day you can come in with a jumper on and it is too hot and the next you come in with a blouse and no jumper and you're freezing … I feel that I am getting continual colds and suffer from general lethargy since I started here. I never suffered from them to the same extent before I came here. The atmosphere is all wrong. There's never any freshness, there's always this feeling of stuffiness irrespective of the temperature. Your skin feels dry. If you wear contact lenses, which I do, your eyes feel dry and gritty all the time.
>
> (Rose, mortgage team member, Finance Bank)

Clearly, at this stage we can hypothesize that the new organizational goals of office-using companies may be significantly undermined by aspects of the working environment.

THE CAUSES OF SBS AND HUMAN RESOURCES

Organizational differences

Most studies of the causes of SBS have, understandably, concentrated on possible technical/structural factors such as ventilation and heating. However, the fact that the incidence and experience of SBS cannot entirely be explained by technical causes is evidenced from the frequent observation that not everyone in a given location experiences the symptoms to the same degree; furthermore these differences, far from being random and a reflection of individual metabolic differences, seem structured or patterned. Several accounts of SBS [21, 23, 24] find both that clerical/secretarial grades suffer more than managerial or technical grades and that women suffer more than men, and similar observations have emerged from our own research in several locations in Glasgow. Clearly, given the feminized nature of much routine white-collar work, grade and gender overlap considerably here (although gender does seem to have a significant effect on its own [26], but more importantly these findings point to work organization as a significant variable in SBS incidence, that is, in the way in which it is experienced and who experiences it. In the Wilson and Hedge survey [21], and in our own interviews, it was observed that management, while noting the health complaints of their staff, often did not seem to share them and often saw nothing wrong with the building. This, of course, may partly be attributable to an awareness of the high cost of building rectification; but it is also a reflection of differences in work organization.

The key concept underlying all current experiments in new forms of organization and management strategy is flexibility: contemporary management thinking dictates that companies and the people who work for them must be increasingly flexible in order to respond to shifts in market demand, technological change and economic environment. The incidence of SBS and similar environmental complaints raises the question of whether the office building can match this flexibility. In many ways the modern office is less flexible than its predecessors as, in a traditional naturally ventilated office, the response to local or temporal variations in conditions was to open or close a window, adjust the heating or turn on a desk lamp. These controls are seldom available to the office workers in the modern office which has been described by Wilson and Hedge [21] as a 'constrained environment', such constraints being located not just in the construction of the building but in the way it is laid out and managed. In line with the philosophy of top-down control over energy use, functions such as heating and ventilation in contemporary office buildings are often in the hands of subcontractors or agents to whom even management must make approaches to alter the settings and operating parameters. Looked at from a human resource perspective there are several things wrong with this, which become apparent if we consider the variations in social activity which the typical commercial building should be able to cope with:

- Differences between organizations: not all organizations are the same or have the same pattern of work activity. For example, government departments may tend to be fairly unchanging or static in their organization, whereas a creative design-based organization will be more fluid, with fewer fixed workstations and considerable variation in the pattern and tempo of work.

- Variations within the organization: even though buildings and their systems are designed with notional occupancy rates and levels of IT usage in mind there is a tendency on the part of building systems designers to assume that everyone in the building will be doing the same sort of thing, rather than an appreciation of the variations in working patterns which typically exist within an organizsation – some people in sedentary static jobs, others more mobile, some working with heat-generating technology (VDUs, photocopiers), others without. We should also note that, following the technological and organizational changes already outlined, we should expect such variations to increase; the 'post-Fordist' office will be a more varied place than the Taylorized office. We can also expect the 'churn rate' – the internal reorganization and relocation of work stations and work groups within the building – to get higher; estimates at present put the churn rate of North American offices at around 30–40% per year but in some locations it is already as high as 70% [26].
- Variations over time: similarly, centralized systems may not be sufficiently responsive to variations in the patterns of working activity over time. Vischer [26] points out that lighting systems for example may be designed to operate under 'worst case' situations (i.e. when it is dark), but that at other times this may result in inappropriate levels of light (especially in relation to VDU use).

At a bank location which we studied in Glasgow (referred to here as Finance Bank) work in the mornings is fairly sedentary and in this period most of the complaints are of low temperatures and draughts. In mid-afternoon the tempo of work picks up quite dramatically as key documents have to be run off and mailed to customers for the following day – staff move about more and there is heavy use of photocopiers and laser printers. In our survey the afternoon period represented the peak in complaints of high temperatures and stuffy air, and was the time most sufferers experienced headaches, tiredness and eye strain. It seems unlikely that the change in the rate of activity caused the quite marked heat gain and deterioration in working conditions, but clearly it may well contribute to them (especially as both photocopiers and laser printers give off ozone as exhaust gas). The question is, is the system flexible enough to cope with these fluctuations? In this particular location it appeared that it was not, or rather the system response time seemed far too slow.

Similarly, changing patterns of work and the increase in the use of office space around the clock will put more pressures on the built environment to support the organization of work as in many cases such new patterns of working time are being introduced into offices designed for more conventional nine to five use.

Buildings, technology and people

If building designers have in the past not paid sufficient attention to the patterns of social activity of occupants, so management may not have paid sufficient attention to the building they occupy beyond its rental, location and total floor space. As will be explored below, the physical environment dropped out of most popular management and social science literature after the Hawthorne experiments in the 1920s, a neglect reflected by its absence in any contemporary management textbook.

SBS suggests that, if that gap was ever justified, it certainly cannot be sustained today. We argue that it is no longer possible to analyse the state of employee commitment, job satisfaction or white-collar occupational health as if these were taking place within a neutral shell. Rather, the built office environment must be seen as a key variable in the analysis of white-collar work which we must add to the more usually considered variables of the technical systems employees use and the patterns of social organization.

The connection between occupational health, the work process and the built working environment becomes a little clearer if we follow the usage of ergonomists in dividing the working environment into the ambient environment (temperature, lighting, noise, air quality) and the proximate environment (the layout of the work station and VDU, furniture, privacy and degree of communication) [27]. To adapt what is referred to in sociological and organizational studies literature as a socio-technical model [28], the total office environment could be said to be composed of :

- the organizational system – management and staff, the work they do, their social relationships;
- the technical system (proximate environment) – the technology used and the constraints/freedoms it places on the work of the staff; and
- the environmental system (ambient environment) – the building and its capacity to deliver consistent and acceptable standards of heat, light, fresh air, space.

As an initial hypothesis we could say that where any two of these systems are out of 'fit' with each other, the result for the office worker can range from dissatisfaction to discomfort to physical illness. For example, a badly designed IT-based job which required VDU work for seven or eight hours a day with poor workstation layout and ergonomics would be an example of a lack of fit between the organizational and technical systems which could result in eye strain, muscular pain and forms of RSI or upper limb disorders. Similarly a building whose poor air quality or level of thermal comfort does not sustain long periods of sedentary work would represent a bad fit between the organizational and environmental systems and could result in the symptoms of SBS and other building-related illness.

From the point of view of the office worker it may seem irrelevant to try to identify which symptoms of ill health are the result of SBS, and which result from stress or prolonged VDU use. We have to see the experience and effects of all these as cumulative. If we try to take a holistic perspective of the office as a combination of building system, organizational system and technical system, we can start to hypothesize that where these overlap – for example someone who is a regular VDU user in a repetitive sedentary job, experiencing stress due to an intensification of workload, in a building with inefficient heating/ventilation and humidity control – then such individuals may be subject to an unpleasant combination of working environmental conditions in which the whole becomes more than the sum of the parts.

THE CONSEQUENCES OF SBS FOR THE HUMAN RESOURCES FUNCTION

If a poorly functioning working environment is experienced as unpleasant and unhealthy by those working in it, their reactions to this can prove costly for the organization. We can discuss these under the headings of individual and collective responses.

Individual responses

Staying off sick

Repeated colds, flu or headaches all contribute to the fact that probably the first indication that a personnel department might have of a problem with the working environment will be a rise in the sickness absence rate. While clearly employees take sick leave for a variety of non-SBS related reasons, a good example of the link which frequently exists between a malfunctioning building and a rising level of sickness absence is given by the case of the Glasgow bank mentioned earlier.

Finance Bank moved from their previous office (Biscay House) in the Glasgow central business district to the newly constructed offices at Clyde Wharf in April 1992. The 300 employees occupied three floors in the building and essentially the same organization of work was transplanted from the old to the new location. After two years' occupation of this state-of-the-art building, a working environment survey of employees asked what employees liked most about it. What was striking was that between one in six and one in seven could find *nothing* at all they liked about working in the building. This was hardly surprising as, in the two-year period between the move and the survey, sickness levels had been rising steadily: in the period immediately prior to the move to Clyde Wharf, the average monthly sickness absence rate was 3.84%; by

1992–93 it had risen to 4.03% and in 1993–94 it was 4.7%, with monthly figures regularly running at around 5% and touching 7.3% in a flu epidemic of November 1993.

It should be noted that the workforce and the technology were the same in the old and the new locations; the only altered variable was the building. The initial low levels are what one might expect in a period of settling into a new building but the rising trend in the period following immediate occupation is certainly consistent with previous findings on SBS. The mean level of sickness absence compared unfavourably with the national annual average absence rates for the finance sector (2.5%) and the UK average for all industries (3.5%) [29, 30].

Productivity

Not all ill-health will result in absence; there may be heavy pressure on the employee to minimize absences (see below), or some SBS symptoms (dry eyes, stuffy nose, lethargy), though debilitating when experienced over a long period, may not seem sufficiently severe to warrant staying off work. However, employees who drag themselves into an unpleasant and disliked workplace will not do much for productivity.

Past estimates of this have either attempted some objective measure of productivity or have recorded what workers feel to be the case. Although the former may seem to be more 'objective', as Raw points out [23], if workers believe the office environment affects their productivity this belief will affect overall commitment and attachment to the organization whether it is justified or not. In the USA, the extensive BOSTI studies in the early 1980s of 4000 office workers over five years demonstrated that office environmental design affects both job satisfaction and job performance. This and other North American studies have led Vischer to conclude that the physical environment of office work may account for a variation of from 5 to 15% in employee productivity [26].

Turnover

If unhealthy environmental conditions persist, or remain unresolved then often things may get so unpleasant that the employee wants to leave, and this will be evidenced in a rise in the company's figure for labour turnover. Because different organizations and industrial sectors may have differing turnover rates, the question of whether this has reached the level of a serious problem will depend on how the suspected 'sick' location compares with others doing similar types of work (the average turnover rates for that particular industry or sector) or other locations in the same company. The most usual method of calculation is

$$\frac{\text{annual leavers in a year}}{\text{average staff in a year}} \times 100$$

usually using a rolling year calculated monthly or quarterly [31]. Although this, as the most usual calculation, allows comparison with other locations, there are problems over the definition of 'leaver' and whether this includes retirees, deaths, dismissals or redundancies. Even if redundancies are excluded, the figures can be affected by the age profile of the workforce and so more useful calculations are based on length of service.

Commitment

As described earlier, HRM strategy is frequently aimed at raising and sustaining the level of employee commitment to the organization. Failure on the part of management to do anything about a disliked and unhealthy working environment is clearly counterproductive. Vischer [26] quotes an example where temporary workers showed less concern with adverse features of the working environment than did permanent employees in the same building; she concluded that the latter were much more critical of lighting, temperature shifts, ventilation and voice privacy because they had more of an investment in the quality of the environment, whereas for the temps the building was simply a convenient short-term source of income.

Altering the environment

In a context of centralized control, it is understandable if individual workers attempt to override the system to assert some sort of localized control over their immediate working environment by bringing in their own fans, heaters or ionizers. This response can often take the form of a low-tech solution to a high-tech environment; for example, when the authors were visiting an extremely new 'intelligent office' in Glasgow during a heat wave they found all the high security-code doors tied back with string, in a desperate attempt to boost air circulation in the sealed and fully mechanical building.

It is symptomatic of the current hierarchical approach to control over working conditions that such attempts are often described in the facilities management literature as 'tampering', and indeed that the use of personal heaters may be banned [26].

Collective responses

Responses such as absence and turnover are both 'individual' responses to an unpleasant work situation – they do not in themselves change that situation but simply withdraw the worker from it. In a unionized work-place a more collective response is more likely, usually based on an attempt to persuade management to accept that a problem exists for their members and to remedy it.

In a national survey of UK trade unions (including both TUC and non-TUC affiliated) in 1993 [25] it was found that most unions with substan-tial white-collar memberships were aware of SBS as a potential health and safety issue; there seemed to be a gathering momentum of union response. Seven unions had discussed SBS at their national conferences, twelve at national executive level and several, especially in the civil ser-vice, had adopted formal national policy on the subject. The civil service union NUCPS was the only one that at that time had a formal agreement with the employer on how to deal with SBS but the civil service joint union body (the Council of Civil Service Unions) had submitted evidence to the Environment Select Committee on Indoor Pollution [32]. Ten unions had sent out some form of information or communicated advice to their members on the subject of SBS.

Negotiation

In the UK there is room for wide disagreement over the quality of the working environment as there are no legislated standards for working conditions (see below); therefore negotiations may focus on gaining con-cessions on agreed maximum and minimum conditions. For example, the banking union BIFU at its 1990 annual conference passed a resolution instructing all levels of the union to negotiate a maximum working tem-perature with employers [25].

At the Glasgow office of a government department (Polar House – built in the late 1980s as an early example of an 'intelligent' building) a mild spring with external temperatures in the mid-60s resulted in inter-nal temperatures in the 85–92°F range – a situation made worse for the workforce by the knowledge that the building was a 'sealed' structure with no opening windows By the summer time the office, in the man-ager's words, looked like Club 18–30 as employees came to work in this 90°F sealed box in shorts and tee-shirts. It was in recounting what it was like to work in this generally harrowing environment – with tempera-tures uncontrollably in the 90s, telephones ringing and non-stop manage-ment pressure in busy periods – that a union representative described the office as a 'stressed sauna', in which the workers went home 'tired, stressed and soaking'.

> It was completely and utterly stressful. We as trade union reps bore the brunt of the dissatisfactions of the members … Tempers were frayed and industrial relations were at an all-time low. There was so much animosity and there was nothing we could do. Management

was sick and tired of us coming in and saying the same thing. We would be going to see them every day so they would lose their tempers. Polar House was a time bomb.

Subsequent heated (in every sense) negotiations covered not just possible remedial measures but also the definition of the situation: union representatives claimed that what workers were experiencing was SBS, but management refused to accept the validity of the term and at first brought in soft drinks and humidifying vegetation which did little to improve the situation. Other remedial measures employed over the period, following continued union representation and the negotiating of a local agreement, included providing stand-alone fans, humidifiers and a huge external chilling unit running the outside length of the building (referred to by all occupants as 'the giant condom').

Permission had to be gained from the Health and Safety Executive to jam open the fire doors to try to improve the air flow; the HSE's condition was that a member of staff had to be deputized to sit and guard the open door. Eventually it was agreed that when the temperature reached 75°F in any area those working there would get a 15-minute break every hour: if it reached 80°F the office closed early. This can be seen as a typical negotiated compromise in an occupational health system with no statutory standards.

Eventually the immediate problems were resolved when the government purchased the building from the developer and spent a considerable sum of money installing opening windows throughout the structure.

Collective action

Where the situation persists or negotiation fails to reach any mutually agreeable solution a further outcome may take the form of collective industrial action. In the current legal climate in the UK this is unlikely to be an actual strike, although at Polar House, as the conditions continued, there were a series of walkouts to which the union officially turned a blind eye because of the legal constraints but tacitly supported. More usually, judging from past examples, action can take the form of everyone going home early when working conditions reach a level of unacceptability.

For example, soon after the Milton Keynes Job Centre opened in 1980, the (all-female) CPSA membership began to suffer from itching eyes, chest problems, coughs, colds, sinusitis and from a general lethargy in the afternoons. Many also experienced a '3 o'clock flush' – a reddening of the face. Sickness levels were the highest of all the job centres in the area, and the staff constantly complained about the office heating, lighting and ventilation to such an extent that, after four years of management inaction, the workforce refused to work in certain areas. Following more fruitless complaints, in the summer of 1985, whenever the office got too hot the union members started going home at 3 o'clock. Management then spent £10 000 on piecemeal repairs which did little to resolve the fundamental problems. Eventually, the union's long-standing demand for an independent survey was accepted by regional management, and the subsequent report identified problems with ventilation, air pollutants, humidity and lighting. Remedial work costing an estimated £30 000 was carried out in order to improve environmental conditions, and sickness levels among office staff subsequently 'dropped dramatically' [33].

Following a long history of minor illnesses at the Inland Revenue offices at St John's House in Bootle, a survey was commissioned from the Building Research Establishment in 1988 after pressure from the civil service unions. Over 40% of staff were found to have left work early at some time due to work-related ailments. Building design and maintenance faults were identified and, the report concluded, by the standards of other buildings in the UK, the premises were 'a sick building'. In 1995, after considering the costs of total refurbishment, the decision was taken to decant the occupants to alternative premises and to demolish the whole building.

HUMAN RESOURCE REMEDIES

Technical and structural remediation of SBS is considered elsewhere in this anthology; but a definite contribution to remediation can be made by personnel or human resources. The problem is that, despite the rhetoric about the HRM function acquiring a more strategic role within the organization which would give it a say in any decisions over premises, the evidence in the UK so far is that this is not happening [34]. As a consequence, the personnel or human resources department will in many cases still only be required to deal with the social symptoms of SBS, such as absenteeism, and unfortunately the policies that first spring to mind here may not remediate the situation but rather may make it worse.

Sickness absence policy

Employee sickness absence as an issue has risen swiftly up the personnel agenda in the UK following the announcement in the October 1993 Budget of the abolition of reimbursement of employers for the cost of employee sick pay from April 1994. In introducing the Statutory Sick Pay Bill, the Department of Social Security stated that it 'would give employers a greater incentive to tackle the high level of absenteeism in British industry'.

The new priority given to sickness absence was reflected in several articles in the management and personnel journals [35, 36] which made a distinction between the 'genuinely ill' and the 60% of all sickness absence which is for periods of less than three days. The inference is that regular patterns of absenteeism of such short duration have no basis in ill health and are therefore a suitable target for action.

This already seems to be resulting in a return to a 'harder' approach to HRM and a stepping up of direct control and explicit or threatened use of sanctions. An example has been set by the government itself, acting as employer, and the Inland Revenue has introduced a 'sickness management' initiative under which staff are paid a bonus of £100 only if they take no more than two days off in six months: the IRSF union's criticism of this innovation included the fact that a pilot version of this scheme was originally planned for the much-plagued St John's House in Bootle (described in the previous section), prior to the decision to demolish it. Elsewhere 'return to work' interviews conducted by management on the day an employee returned from sickness have been claimed to lower absence rates, while the Rover Company operates a points system based on the Bradford Factor, which penalizes employees with frequent short-term rather than long-term absences [29]. In our own research we have encountered a building from which several employees had been dismissed for persistent absence, which the employees and their union attributed to SBS-induced ill health.

While one can see that such policies and approaches may be tempting as an immediate management response to rising absence costs, for those workers suffering from the symptoms of SBS, and who are therefore likely to build up a record of regular (and short-term) absences, the result will probably be to pressurize them into coming to work when ill. This, rather than curing the problem, will probably worsen it as not only can minor ailments develop into major building-related diseases through increased sensitization, but as noted in the previous section, the whole emphasis of this policy runs counter to the longer-term HRM goal of raising employee commitment.

If staff already feel committed to the workplace, one would not expect them to stay away more than they can help. For example, in a survey of a government department located in Glasgow, 83% of respondents said they sometimes came to work when feeling ill, while another 8% said this was 'often' the case and the rank order of reasons given for coming to work when feeling ill was, in descending frequency: 'not ill enough to stay off', 'backlog, pressure of work', 'commitment to the department', 'management pressure', and 'no option (temporary workers)'. If such staff are then penalized when they become more seriously ill this level of commitment is unlikely to be maintained.

Health and safety legislation

The legal issues surrounding SBS are examined more fully elsewhere in this collection; but it is worth mentioning those aspects of health and safety legislation which bear on HR policy. Although section 2(i) of the Health and Safety at Work Act imposes the duty on the employer 'to ensure as far as is reasonably practicable the health safety and welfare of all his (sic) employees' [37] until recently the law on occupationally induced ill health has most definitely favoured the employer. Under the Employment Protection (Consolidation) Act, 1978, it is perfectly legal (or 'fair') to sack an employee for persistent sickness if it can be shown that the employee is unfit to perform the work for which they were employed (as mentioned above, we have encountered instances of this in our research in cases of SBS). Unlike the situation in the USA, however, it is comparatively rare for an employee to be able to successfully sue the employer for ill-health compensation. A recent exception was the successful case in 1995 of *Walker* v. *Northumberland County Council*, where a former employee successfully claimed that he had been placed in a stressful work situation despite the employer knowing that the same work had contributed to an earlier nervous breakdown [37]. This has been interpreted as an admission that the employer has a duty of care to staff not to cause them mental ill health.

Probably the most important recent legislation relating to SBS is that arising from the 1992 European Directives on Health and Safety. While these mainly codified existing practice in the member states, their most important feature is that they place the prime duty on the employer to assess and remedy all possible risks to workers' health. The Management Regulations spell out this general duty while the Workplace (Health Safety and Welfare) Regulations and their Approved Code of Practice give very specific guidance on heating, lighting and ventilation, cleanliness and space dimensions [38]. In the Workplace Regulations there is specific mention of the need to maintain the workplace and its equipment in an efficient state; efficiency is specifically defined as relating to health, safety and welfare and not productivity or economy. There is particular mention of the need to clean air conditioning systems and mechanical ventilation systems to maintain an adequate level of hygiene. Workers and their representatives must be kept informed of all measures taken to improve health and safety in the workplace.

In general, however, while health and safety legislation may be sound in principle it is often weak in its specific wording. For example, fresh or purified air must be supplied in 'sufficient' quantities and stale air should be removed at a 'reasonable' rate and clearly such terms are open to disputed interpretation.

Occupational health service

We have argued earlier that organizations can no longer ignore the quality of the working environment, and the same argument extends to the wider issue of employee health. If the goals of HRM are taken seriously by the organization they will include the formulation of a company health policy and already companies with an HRM strategy increasingly provide occupational health programmes which cover nutrition, smoking advice, fitness programmes and the like. However, in relation to SBS these individually aimed 'lifestyle' programmes cannot be seen as a substitute for identifying and eliminating unhealthy working conditions, just as 'stress counselling' is no substitute for designing less stressful jobs. However, many workplaces still have no occupational health service (OHS) as there is no legal requirement to provide one and the UK has not ratified the ILO Convention on OHS.

Where company OH services do exist, the way in which they have been used in the past was revealed by a survey in the late 1980s which found that many employees see the OHS almost as the medical arm of management charged with 'checking up on workers' if they have been off sick for too long [39]. The House of Lords Committee recommendation in 1983 that there should be trade union representation on OHS committees together with engineers and technologists seems to have been widely ignored, as has the HSE's Action Programme of guidance on OHS in 1986.

Training

The staff responsible for controlling the building management system, HVAC system or equivalent may be employees of the building owners, factors or managing company rather than direct employees of the tenant or occupying company. Either way, it is important to determine whether they have had sufficient training to operate what, in contemporary offices, are increasingly sophisticated control systems. In our own research we have found more than one location where the responsibility for fine tuning the controlling software was in the hands of staff without the necessary computer-based training: in one case in the hands of a traditional electromechanically trained service engineer, in another in the hands of security guards on the main gate. It is not surprising that when occupying organizations rang down to complain about temperature variation they got very little noticeable satisfaction and the mood in each location had rapidly become a fatalistic 'It seems like there is nothing you can do'.

This is a good case of a bad 'fit' between the environmental and organizational systems. Large amounts of money had been spent on constructing the building with a state-of-the-art control system, the tenant was paying a substantial rental for this city centre property and yet, presumably, someone had economized on an adequate training and induction programme for those staff responsible for seeing that the system delivered what its manufacturers claimed it would deliver.

Involve the workforce

One of the first things management will need to know is whether the complaints it has been receiving (whether concerning the building or repeated ill-health symptoms) come from the same few individuals (and if so whether this is due to their sharing a common location) or whether it is a general problem distributed throughout the building and the organization. The most effective way to do this is to conduct an employee survey. As the HSE recommends, 'The easiest way to highlight problems is ... to ask the staff themselves. They will generally be the first to know about problems with temperature control, lighting, noise levels, stuffiness, fumes and tobacco smoke. You may also learn something about people's attitudes to their work' [38].

If there is an existing health and safety committee this is an obvious forum through which to conduct such an investigation; but if there is not, then some sort of joint consultation mechanism on the working environment should be considered.

DISCUSSION

The phenomenon of sick building syndrome raises several major theoretical issues which affect our understanding of people and organizations. These involve the nature of the relationship between the physical work environment and the performance of work – motivation, output and productivity – and the general question of occupational health and the different approaches to this in differing employment systems. Lastly we need to consider the desirability of increasing the degree of employee involvement in both the control of the immediate working environment and in the actual design and planning of such environments.

The physical environment and work

In the first part of the century the early industrial psychologists paid a great deal of attention to the physical environment in the workplace as part of the study of the phenomenon of fatigue. The 'human factors' school of researchers who followed the British psychologist C.S. Myers argued that F.W. Taylor's view of the worker as a money-motivated automaton was too crude and that a whole range of influences on worker behaviour had to be taken into consideration, including the physical characteristics of the working environment such as standards of ventilation, heat and lighting [40].

Ironically it was a continuation of this approach which lay behind the early Hawthorne studies, which in turn signalled the end of this brief period of interest in the physical environment by the social sciences. In 1924–25 the Western Electric Company had commenced its own experiments on working conditions at its Hawthorne plant near Chicago, the first series of which concentrated on levels of lighting. In the best-known case, lighting levels were raised and lowered in both an experimental work group and a control group to see if there was any effect on productivity. The results were perplexing – output continued to rise in both groups, even when light levels in the experimental group were reduced to minimal levels.

In the analysis of these and later experiments on work group behaviour, subsequently publicized by the Harvard psychologist Elton Mayo, the conclusion was reached that the reason output rose was the very fact that the workers knew they were being observed (the so-called Hawthorne effect) and collectively altered their level of effort. This observation became one of the foundations of the subsequent human relations emphasis on the importance of social groups, social organization and group norms in the workplace as being more powerful determinants of employee behaviour than physical conditions.

The result of this change in emphasis, however, was that it was assumed that physical conditions were not worth studying. Actually, all that the lighting experiments showed was the consequence of a badly designed experiment – it did not actually discover what the effects of different lighting levels were because it failed to screen out other determinants of behaviour.

In 1959 Herzberg included work conditions as one of his 'hygiene' factors in his two-factor theory of motivation. While things intrinsic to the job, like achievement and responsibility, were seen by Herzberg to be motivators for the worker, other extrinsic hygiene factors – aspects of the work environment rather than the work itself – such as company policy and supervision and working conditions, were demotivators if not present at an acceptable level. (The term 'hygiene' is an indicator that Herzberg was influenced by a medical model based on the analogy of health, in which the absence of certain things (such as vitamins) makes us unhealthy but their presence over a given level will not make us significantly more healthy.)

The trouble with all these psychological and human relations approaches is that they still assume that the employee is a sort of flat battery which various factors in the work situation will charge up, leading to varying levels of 'motivation'. They also assume that once motivated the goals of the employee will be the same as the goals of the organization. None of these approaches considers the possibility that the structure and balance of hierarchy and authority relations in the workplace are more likely to affect the outcome of particular working situations than individual levels of motivation on their own. We have seen from the examples quoted earlier that health and the working environment are often contested issues, with management and employees approaching them with differing priorities. We can therefore repeat our contention that the physical environment must be reintroduced to any analysis of office work but we should now add to our three variables of environment, technology and organization the realization that the outcome of health issues will depend on the nature of the employment relations and human resource practices. We thus need to consider the question of occupational health more deeply and then to put it into the context of employment relations.

Occupational health

While the International Labour Organization describes health as 'a state of complete physical, mental and social well-being and not merely the absence of disease or deformity' [41], past British approaches to health and safety, while good in themselves, have been mainly concerned with safety – the avoidance of accidents by guarding machinery, and the proper use and handling of toxic or dangerous substances. This traditional 'machine-minding' approach not only falls short of the wider view of occupational health indicated by the ILO definition, but also ignores some of the organizational dimensions to occupational health and safety which we had indicated above; for example, the level of accidents in early nineteenth-century textile mills was only partly due to unguarded machinery – it was also due to long working hours, inadequate earnings, inadequate training, and the use of child labour [41].

The limitations of this approach with regard to SBS are first that, in this 'industrial' view of safety, the office may seem non-problematic as it has no dangerous machines, moving parts nor toxic processes. Secondly, the notion of employee health as an issue has until recently been absent from UK discussions of health and safety policy; it is only with the eventual acceptance of the dangers to health posed by the wrong use of information technology that the idea of white-collar health has come to be gradually accepted as an issue.

Perhaps the biggest potential pressure on employees to rethink the conventional approach to employee health and safety has come from the 'six-pack' Health and Safety Regulations of 1993, which have their origins in the directives emanating from the Action Programme of the European Social Charter of 1987. As indicated in the previous section, these are interesting in that they take a more 'European' approach of making the employer responsible for assessing risks to the health of employees and emphasize that environmental discomfort should not be seen as some sort of inevitable price to pay for work of any type.

In northern Europe and Scandinavia there has been a much longer tradition of concern with occupational health. The 1977 Swedish Working Environment Act was drawn up with complete involvement of the unions and over the years the scope of the Act has been extended, through practice in the workplace, from accident prevention to the promotion of mental health and job satisfaction. The Act is a 'paving' or enabling act which has to be supported by statutory instruments and collective agreements. Responsibility for the occupational environment is placed primarily on the employer and these responsibilities are essentially of a preventative, rather than a remedial nature. The Act is based on the idea that a good occupational environment must be guaranteed by cooperation between employer and employees at workplace level and so it requires workers to be involved in the planning and design of both the workplace and work tasks. Safety stewards and safety committees are part of their companies' decision-making process, with jurisdiction over all aspects of the design of work and the working environment.

The industrial relations of health

It seems reasonable to conclude from the above that conflictual employment systems (such as those in the UK or USA) give rise to a health and safety approach characterized by seeing the working environment as a given factor: any negative influences must be minimized after occupation through the negotiation of 'safeguards'. Health here thus becomes a bargaining issue and a potential area for a conflict of interests or priorities – health versus energy saving, for example.

In employment systems with a more consensual or cooperative tradition, while wages and hours may be strongly contested, the quality of the working environment is more likely to be seen as an area of common interest on which agreement can be reached. This and the tradition of employee involvement through works councils and similar participative structures, results in a higher level of employee involvement in the creation of a healthy working environment rather than simply minimizing unpleasant conditions which have already been created.

User involvement in building design

In Sweden, Germany and Holland we find a much greater tradition of buildings being designed specifically for the user organization. As we have seen in the case of Sweden, workers have the statutory right to be involved with employer's plans for changes to the workplace, resulting in negotiations commencing before a building is designed. Legislation such as the Norwegian Working Environment Act, which demands a 'fully satisfactory work environment' to be achieved through joint working environment improvement programmes, means that employers are more likely to take the social effects of building design into account in the design stages rather than treat them as an afterthought. In Norway the owner has to specify the characteristics of the building he wishes to buy and after consulting with the builders and technical consultants the specifications (including test results) are shown to the Labour Inspectorate.

Duffy [7] suggests that the consequences for the working environment of an increasing emphasis on employee participation will be an enhanced importance for access to natural daylight, natural ventilation and opening windows (a 'domestic' model) together with an emphasis on the amenity value of an attractive site and the quality of the environment in terms of employee health (absence of SBS, no smoking policies).

It is not surprising that those buildings hailed by architects such as Duffy as examples of best practice tend to be found in these environments. In Holland there has been a great deal of attention given to the desirability of buildings with openable windows, the much-lauded NMB (now ING) Bank building in Amsterdam was built to be both energy efficient while ensuring that no worker was far from a window. In Sweden the move away from deep open plan offices has produced buildings like the SAS building in Stockholm with a linear design, small personal office spaces and open public 'meeting spaces' [42].

Localized control

Social factors do not cause SBS; but the fact that work organization and job design factors contribute to the way it is experienced is acknowledged by the Health and Safety Executive in its latest guide for managers on handling SBS. The HSE states: 'These physical and environmental problems can be exacerbated by organisational factors. For example, lack of personal control over working conditions in open plan offices, or lack of work variation, can reduce decision making powers and hence job satisfaction' [38]. The HSE's argument here is that dissatisfaction with the job leads to more complaints about the working environment. However, it could be argued, using our earlier three-factor model of 'fit', that the reverse is true and that the more employees try to complete tasks and meet goals in an unsupportive environment which is making them ill, the less satisfaction will be found in other aspects of the job (including the pay) and thus the less likely the organization is to sustain the levels of commitment to corporate goals that HRM policies are designed to secure.

The absence of localized control over the working environment by those who work in it has been noted as a significant variable in several analyses of SBS. Wilson and Hedge [21] concluded from their Office Environment Survey that dissatisfaction appeared to be exacerbated by the fact that many office buildings were designed in such a way that the occupants had little ability to control their environment and also that where control is reduced people do not just experience discomfort but can regularly experience ill health.

Differences in the ability to control the environment coincide with some of the observed organizational differences in the incidence of SBS in a given building. Clerical grade staff are more likely to be located in open plan offices which by their nature offer the individual very little autonomy in setting environmental levels; secondly their work is sedentary and locationally fixed (compared to managerial grades) entailing less freedom to move to a different environment; and, thirdly, repetitive, visually demanding, jobs have more stringent ergonomic requirements.

However, improving control is not the same as giving employees the impression of control through such dubious strategies as – in a number of recorded cases – fitting 'dummy' or placebo thermostats 'to keep users happy' [21] while allowing centralized control over energy use to continue.

CONCLUSION

Syndromes are essentially bundles or recurring patterns of symptoms for which at present we do not know the cause. Like other current syndromes (Gulf War, ME) SBS is an elusive phenomenon, and vulnerable to being dismissed by the ignorant. Yet for the many thousands of office employees and staff who put up with its effects on a daily basis, it can transform the humdrum business of going to work into something approaching torture.

The problems of analysis and remediation have been compounded by a tradition of professional segmentation in perspectives and priorities. Up to now analysis of what goes on in the office has been fragmented between social/organizational scientists (job content, work organization, technical change), psychologists, occupational health practitioners and trade unions (stress, ergonomics, office safety, RSI), and architects and building designers (SBS and internal environment, intelligent buildings). We would argue that the creation of healthy working environments requires a holistic multidisciplinary approach.

Designing buildings with enhanced worker control over local environmental conditions should not be impossible, and has indeed been seen as a priority in new buildings in Scandinavia and the Netherlands. While there is by no means an absence of building-related ill-health in these countries, it is clear that the earlier the stage at which employees and their representatives can be involved in the design of the workplace, both structurally and organizationally, the more chance there is of creating a work space that is both effective, pleasant to work in, and conducive to health.

This seems to be less possible in those office property markets (such as

North America and the UK) where offices are often constructed speculatively and then sold or leased to the end-user organization. Here not even the management has much of a say in the design of the building and often modifies the design once the building is occupied (for example by erecting partitions in a space designed as open plan).

The chances of the best in architectural practice being combined with the most effective work organizational practices may require a more structured social and legal framework than that provided by the market alone.

The research upon which this chapter is based was grant-aided by the Leverhulme Trust.

REFERENCES

1. Markus, T. (1993) *Buildings and Power: Freedom and Control in the Origins of Modern Building Types*, Routledge, London.
2. Cowan, P., Fine, D., Ireland, J., Jordan, C., Mercer, D., Sears, A. (1969) *The Office: A Facet of Urban Growth*, Heinemann, London.
3. Duffy, F. (1980) Office buildings and organisational change', in *Buildings and Society: Essays on the Social Development of the Built Environment* (ed. A. King), Routledge, London, pp. 255–82.
4. Lockwood, D., (1958) *The Black Coated Worker*, Unwin, London.
5. Moor, N. (1979) 'The contribution and influence of office developers and their companies on the location and growth of office activities', in *Spatial Patterns of Office Growth and Location* (ed. P.W. Danils), John Wiley, Chichester.
6. Castells, M. (1989) *The Informational City: Information, Technology, Economic Restructuring and the Urban-Regional Process*, Basil Blackwell, Oxford.
7. Duffy, F., Laing, A., Crisp, V. (eds) (1993) *The Responsible Workplace*, Butterworth, Oxford.
8. Baran, B. (1988) Office automation and womens' work: the technological transformation of the insurance industry, in *On Work: Historical, Comparative and Theoretical Approaches* (ed. R. Pahl), Blackwell, Oxford, pp. 684–706.
9. Nelson, K. (1988) Labour demand, labour supply and the subordination of low-wage office work, in *Production, Work, Territory: the Geographical Anatomy of Industrial Capitalism* (eds A. Scott and M. Storper), Unwin Hyman, London, pp. 149–71.
10. Baldry, C. (1988) *Computers, Jobs and Skills: the Industrial Relations of Technological Change*, Plenum Press, New York.
11. Handy, C. (1984) *The Future of Work*, Basil Blackwell, Oxford.
12. Laing, A. (1993) Changing business: post-Fordism and the workplace, in Duffy et al. (1993), pp. 33–43.
13. Collinson, D. (1993) Introducing on-line processing: conflicting human resource policies in insurance', in *Human Resource Management and Technical Change* (ed. J. Clark) Sage, London, pp. 155–74.
14. Lyon, D. (1994) *The Electronic Eye: the Rise of Surveillance Society*, Polity Press, Cambridge.
15. Craig, M. and Phillips, E. (1991) *The Office Workers' Survival Handbook*, 2nd edn, The Womens Press, London.
16. Peters, T. and Waterman, R. (1982) *In Search of Excellence*, Harper Row, New York.

17. Storey, J. (1992) *Developments in the Management of Human Resources*, Blackwell, Oxford.
18. Walton, R. (1990) From control to commitment in the workplace, in *Managing People Not Personnel*, Harvard Business Review, Boston, MA, pp. 89–106.
19. Cunningham, I., Hyman, J., and Baldry, C. (1996) Empowerment: the power to do what?, *Industrial Relations Journal* (forthcoming).
20. Atkin, B. (1988) *Intelligent Buildings*, Kogan Page, London.
21. Wilson, S. and Hedge, A. (1987) *The Office Environment Survey: a Study of Sick Building Syndrome*, Building Use Studies, London.
22. Sykes, J. (1988) *Sick Building Syndrome: a Review*, HSE Specialist Inspector Reports No. 10, Health and Safety Executive, London.
23. Raw, G. (1992) *Sick Building Syndrome: a Review of the Evidence on Causes and Solutions*, HSE Contract Research Report No. 42, Building Research Establishment/HMSO, Watford.
24. Stenberg, B. (1994) *Office Illness: the Worker, the Work and the Workplace*, University of Umeå, Umeå.
25. Bain, P. and Baldry, C. (1995) Sickness and control in the office: the sick building syndrome, *New Technology, Work and Employment*, **10** (1).
26. Vischer, J. (1989) *Environmental Quality in Offices*, Van Nostrand Reinhold, New York.
27. Dainoff, M.J. and Dainoff, M.H. (1987) *A Manager's Guide to Ergonomics in the Electronic Office*, John Wiley and Sons, Chichester.
28. Trist, E. *et al.* (1963) *Organisational Choice*, Tavistock Institute of Human Relations, London.
29. Incomes Data Services (1994) *Absence and Sick Pay Policies*, IDS Study 556, June 1994.
30. Industrial Society (1994) *Managing Attendance*, Managing Best Practice No 6.
31. Incomes Data Services (1995) *Labour Turnover*, IDS Study.
32. Council of Civil Service Unions, *Evidence to House of Commons Select Committee on the Environment*, London, CCSU, January–February 1991.
33. London Hazards Centre (1990) *Sick Building Syndrome: Causes, Effects and Control*, London Hazards Centre Trust, London.
34. Legge, K. (1995) HRM: rhetoric, reality and hidden agendas, in *Human Resource Management. A Critical Text* (ed. J. Storey), Routledge, London.
35. Norris, P. and Newlyn, J. (1994) Are we really sick?, *Local Government Management*, Spring.
36. Watkins, S. (1994) 'Sick excuses for swanning around, *Personnel Management*, April.
37. Goldman, L. (1995) *Sick Building Syndrome: Causes, Cures and Costs*, Technical Communications, Hitchin.
38. Health and Safety Executive (1995) *How to Deal with SBS: Sick Building Syndrome – Guidance for Employers, Building Owners and Building Managers*, HSE Books, London.
39. Labour Research Department (1989) *Workplace Health: a Trade Unionists' Guide*, LRD, London.
40. Rose, M. (1988) *Industrial Behaviour*, 2nd edn, Penguin.
41. Quinlan, M. (ed.) (1993) *Work and Health: The Origins, Management and Regulation of Occupational Illness*, Macmillan, Melbourne.
42. Doxtater, D. (1994) *Architecture: Ritual Practice and Co-Determination in the Swedish Office*, Avebury, Aldershot.

CHAPTER 8

Legal issues

John Clark

INTRODUCTION

For many years there has been liability upon the owner and occupier of premises in England and Wales in relation to persons working in those premises and also visitors to the premises. Liability to visitors can be found under the Occupiers Liability Act 1957 and liability to trespassers under the Occupiers Liability Act 1984. This legislation does not apply in Scotland; the contents of this chapter relate to the law in England and Wales.

While premises may be visited by persons on a daily basis who may thereby come into contact with SBS the likelihood of persons in buildings becoming affected by SBS relates primarily to employees; this chapter will concentrate upon the liability of employers in relation to their employees if such employees become affected by SBS.

The liability of an employer can be based upon one of two areas of English law, namely:

- breach of statutory duty; and/or
- negligence.

BREACH OF STATUTORY DUTY

In 1972 the Robens Committee reported on safety and health at work. Following the report of the Robens Committee [1] the Health and Safety at Work Act 1974 was produced. Prior to this Act there had been a complex series of piecemeal statutes and regulations dealing with particular types of workplace risk. Many areas were not covered and most statute law in existence was as a result of reaction to an accident or disaster that had taken place. The Health and Safety at Work Act attempted to set down general principles and duties upon employers, the self-employed and persons otherwise in control of premises in order to ensure safety generally to not only employees, but all persons affected by work activities.

HEALTH AND SAFETY AT WORK ACT 1974

2(1) It shall be the duty of every employer to ensure, so far as is reasonably practicable, the health, safety and welfare at work of all his employees.

2(2) Without prejudice to the generality of an employer's duty under the preceding subsections, the matters to which the duty extends includes in particular:-

a) the provision and maintenance of plant and systems of work that are, so far as is reasonably practicable, safe and without risks to health ...

b) the provision and maintenance of a working environment for his employees, that is, so far as is reasonably practicable, safe, without risks to health, and adequate as regards facilities and arrangements *for* their welfare at work.

The test of what is 'reasonably practicable' is not merely what measures were physically or financially possible. The degree of risk must be weighed against the sacrifice involved. If the court finds that the sacrifice is disproportionately heavy then it is likely to conclude that the measures required are not 'reasonably practicable' [2].

Effect in relation to SBS

The Act places emphasis upon accident prevention rather than, as previously, response to events which have already taken place. General duties are placed on employers and others to ensure so far as is reasonably practicable the health, safety and welfare of all employees. Under section 2.2 this is expanded in particular to include the working environment. If, therefore, the environment is such as to produce symptoms of SBS the provisions of the Health and Safety at Work Act would be relevant. However, most importantly the duties of employers do not support civil liability, but rather are the basis for criminal prosecution and administrative enforcement by the Health and Safety Executive (HSE).

The previous law had become outmoded and related to nineteenth-century industry. The Health and Safety at Work Act provided for modern health and safety Regulations to be produced and to be enforceable [3]. In particular, it is provided that Regulations so produced will support civil liability unless provided otherwise, and would be criminally enforceable [3].

Duties provided for under the Act itself do not attract civil liability; duties arising under Regulations made pursuant to the Act do attract civil liability unless provided otherwise.

General principles

Civil liability can arise only when a plaintiff claimant can establish that Parliament intended that the breach of any particular relevant statutory duty shall be actionable by an individual harmed by the breach. It must be shown that Parliament intended to confer a private right of action 'pending in damages' [4]. Provisions designed to protect a class of individuals generally will not be sufficient.

The courts have shown themselves ready to infer a right of action for breach of those statutory provisions which are designed to ensure the personal safety of persons, particularly employees. A plaintiff must show:

1. that the injury suffered is within the ambit of the statute;
2. the statutory duty imposes a liability for civil action;
3. that the statutory duty has not been fulfilled; and
4. that the breach of duty has caused the injury suffered.

EUROPE

The Treaty of Rome signed on 25 March 1957 formed the basis of what was then known as the European Economic Community. The UK joined the EEC on 1 January 1973. The EEC has now become the European Community comprising 15 member states.

The Single European Act 1986 came into force on 1 July 1987 and provided for a Single European Market to come into existence by the end of 1992. One of the purposes of the Single European Market was to promote harmonization of standards between member states, in particular relating to environmental health and safety and consumer protection.

On 1 November 1993 the Treaty of European Union (the Maastricht Treaty) came into force; among other concepts it promoted the European Union comprising common foreign and security policy, the current European Community and an intergovernmental home affairs and justice policy. Furthermore, under the concept of 'subsidiarity' decisions by EC institutions were to be limited to those matters which were strictly necessary and wherever possible decisions were to be made by national, regional or local government.

European legislation

Primary legislation is set out in the form of Treaty provisions together with their protocols and annexes. Some provisions are required to be verified by the legal system of the member state whereas others are directly effective. Secondary legislation takes the form of regulations, directives and decisions. This legislation is within the framework of the Treaty provisions and may be found to be invalid if in conflict with Treaty provisions.

Regulations

There is no discretion on member states as to whether or not regulations of the EC may come into force. It is compulsory for member states to implement regulations. Article 189 of the Treaty of Rome provides 'a Regulation shall have general application. It shall be binding in its entirety and directly applicable in all member states.'

Directives

A member state is bound to implement a directive of the EC, but the choice of method of implementation is left to its discretion. Article 189 also provides 'a Directive shall be binding, as to the result to be achieved, upon each member state to which it is addressed, but shall leave to the National Authorities the choice of form and methods.'

As a result of a series of decisions from the European Court of Justice it is now established that private individuals can obtain directly enforceable rights under directives. This may as yet be limited to claims against the state or an employer in the public sector. However, in more general terms directives are particularly relevant when interpreting regulations. There may be many instances where the wording of a regulation does not coincide with the wording of a directive. Similarly construction will need to be placed upon the wording of UK regulations. It is clearly established that UK law which has been introduced to meet the UK's Community obligations must be construed in accordance with the applicable community law. A directive may override national law if inconsistent with it.

While it has been clear law that an employee in the public sector can take action where an employer fails to meet the standards of a Directive, there has more recently been a widening of this principle to employees in the private sector. Under the decision of the European Court of Justice in *Francovich* v. *Italian Republic* [5] it is established that there is a right to compensation for an individual who has suffered loss or damage as a result of the state's failure to act. This claim should be brought in the courts of the member state. There are three conditions, namely:

1. The directive must confer rights for the benefit of individuals.
2. The content of these rights may be determined by reference to the provisions of the directive.
3. There must be a causal link between the breach of the obligation of the state and the damage suffered by the person affected.

The *Francovich* case concerned a complete failure by the Italian government to implement the terms of the directive. However, the case appears to have removed the apparent distinction between employees in the private sector and those in the state sector.

If the regulations produced by the UK government do not fulfil the requirements of the European directive it would appear likely following *Francovich* that an individual will have a remedy in the English courts against his employer for failure to comply with the directive.

UK REGULATIONS

The management of Health and Safety at Work Regulations 1992

Background

These regulations implement EC Directive 89/391/EEC and are made under the provisions of the Health and Safety at Work Act 1974. The regulations came into effect on 1 January 1993.

Under Regulation 15 a breach of duty imposed by the regulations does not confer a right of action in any civil proceedings. The regulations are admissible in criminal proceedings. It is clearly established that member states must provide realistic remedies in domestic law in order to make Community rights a reality. There would seem little use in having these regulations if they cannot be enforced other than through criminal proceedings. It remains a matter for the European Commission to take up with the UK government regarding the unenforcibility of the Management of Health and Safety at Work Regulations 1992. If it can be shown that an employee has suffered through breach of these regulations there is no reason in principle why the employee should not be able to take action against the state for breach of the directive in particular in failing to provide realistic remedies in the UK courts to implement the regulations.

Applicability to SBS

The main theme of these regulations is the requirement on all employers to carry out a risk assessment under Regulation 3. This must be suitable and sufficient and consider the health and safety of all employees for the purpose of assessing those measures which need to be taken to comply with the relevant statutory provisions. The statutory provisions include the general duties under the 1974 Act as well as many more specific provisions.

The risk assessment must be reviewed if there is any reason to believe it is no longer valid or if there has been any significant change in the matters to which it relates. The significant findings of the assessment have to be recorded if the employer employs five or more employees.

Employers are obliged to give effect to arrangements for the effective planning, organization, control, monitoring and review of the measures needed to be taken following the risk assessment. The arrangements must be in force where five or more persons are employed. Furthermore, an employer must provide employees with appropriate health surveillance.

Thus it follows that an employer of, for example, more than five office workers must carry out a risk assessment of all health risks that those office workers may meet. The assessment will have to consider the possibility of SBS. The risk assessment must continue to be monitored and if there is any sign of SBS recorded arrangements must be made for the future control of SBS. Furthermore, the employer must make arrangements for surveillance of the health of his or her employees. Failure to do so under these regulations cannot produce a civil remedy for breach of statutory duty, but there are possibilities of a claim against the government for failure to implement the directive and furthermore failure to comply with these regulations may be a ground for a claim in negligence (see below).

The Workplace (Health, Safety and Welfare) Regulations 1992

Background

These regulations implement Directive 89/654/EEC and again are made under the Health and Safety at Work Act 1974. They came into force on 1 January 1993. Although the regulations initially applied to any new workplace or new part of a workplace and transitional provisions were included in the regulations it is now the case that as from 1 January 1996 the Regulations will apply to all work places in existence. Most importantly, the regulations do create civil liability.

Applicability to SBS

The workplace, equipment, devices and systems should be maintained in an efficient state, in efficient working order and in good repair. This is provided under the Code of Practice to apply to matters of health and safety rather than to production or working.

There must be effective and suitable provision for ventilation of every enclosed workplace by a sufficient quantity of fresh or purified air. Plant used for ventilation must have an effective device to give visible or audible warning of any failure of the plant where necessary for reasons of health or safety.

There must be suitable and sufficient lighting, so far as is reasonably practicable by natural light.

Under Regulations 20 to 25 it is noteworthy that restrooms and rest areas must include suitable arrangements to protect non-smokers from discomfort caused by tobacco smoke.

Thus it might be seen that if SBS can be shown to be linked to deficiencies of ventilation or lighting which failed to comply with the requirements of these regulations an individual may have a sound claim in damages against his employer for breach of the employer's statutory duty, in particular arising from failure to comply with these regulations.

Negligence

The liability of an employer to his employee at common law is part of the general law of negligence. In order to establish a claim a plaintiff must show that:

1. The defendant owed the plaintiff a duty of care.
2. The defendant was in breach of that duty.
3. As a result of that breach the plaintiff suffered damage.

The Royal Commission on Civil Liability and Compensation for Personal Injury 1978 considered the question of liabilities for employers in relation to employees who might contract a disease at work. It was felt that the onus of proof must remain with the plaintiff employee.

Duty of care

An employer is bound to take reasonable care for the safety of his employees, and in particular not to carry on operations as not to subject those employed by the employer to unnecessary risk (Lord Herschell in *Smith* v. *Baker* (1891) AC325HL). This duty has been separated into competent staff, adequate materials and a proper system and effective supervision. However, this division is more for convenience and the courts will look at the general duty of an employer to take reasonable care of his employees.

The duty of the employer extends only to protecting the employee against personal injury. However, personal injury can include psychiatric injury. Personal injury will obviously include the symptoms of SBS. The place where the duty of care is owed will ordinarily be the premises of the employer.

Standard of care

It is expected that an employer will exercise the standard of care of an ordinarily prudent employer. In order for a court to determine whether an employer has fallen below the required standard of care consideration will be given to the foreseeability of the existence of a risk assessment of the magnitude of the risk and devising reasonable precautions. The risk assessment referred to under the regulations becomes most significant. If a risk assessment has not been carried out the employer is likely to have been found to have fallen below the standard of care and to have failed to foresee the existence of a risk. If the employer has carried out a risk assessment and considered all its implications and taken appropriate precautions to cover matters thereby arising then an employer is likely to have reached the appropriate standard of care.

The vast majority of the law in relation to negligence concerning the liability of employers relates to accidents at work rather than the contraction of industrial disease. However, once it is known to an employer that a particular disease such as SBS is present in a building the employer must take immediate steps to remedy the problem, failing which the employer may be liable in negligence. Great relevance therefore attaches to previous complaints from employees of similar symptoms. Conversely the absence of previous complaints, although material, is by no means conclusive.

Breach of duty

The proof of negligence lies on the plaintiff. He must show that the employer's breach of duty has materially contributed to the damage. The plaintiff in an SBS claim will call expert surveying/engineering evidence to establish the cause of SBS in the building concerned and will furthermore call expert medical evidence to show that the symptoms were the result of SBS in the particular building. The plaintiff must not only prove that the defendant was negligent, but also that the negligence was the cause of the injuries complained of. There may be two or more causes of the employee suffering from SBS. The employer cannot excuse himself by pointing to the other cause. It is enough to show that the conditions of work were one of the causes of the personal injury suffered by the employee.

Lord Reid in the House of Lords has set out the following test to be applied in these cases: 'when a man who has not previously suffered from a disease contracts that disease after being subject to conditions likely to cause it, and when he shows that it starts in a way typical of disease caused by such conditions, he establishes a prima facie presumption that his disease was caused by those conditions.'

One aspect of the duty of an employer to take reasonable care for the safety of his employees is the duty to take reasonable care to provide a safe place of work. Each case will depend on the circumstances. Most case law relates to keeping the workplace free of obstructions which might cause accidents. Nevertheless, this aspect of the employer's duty will include the duty of an employer to ensure that there are safe systems of ventilation and lighting such as might prevent the employee from contracting SBS.

Furthermore, an employee is under a duty to use reasonable care to provide safe plant and appliances. The plant in particular would include the ventilation system. If it could be demonstrated that the ventilation system circulated air which contaminated the building causing the symptoms of SBS then it might be possible to show that the employer had fallen below the required standard of care for the purposes of negligence.

Damages

If an employee can show duty of care, breach of duty and damage the employee will be entitled to damages as compensation for the loss and personal injury suffered. If time has been lost from work due to an employee contracting an industrial disease that employee would be entitled to receive loss of earnings and the payment of any other expenditure. This should be relatively easy to quantify in the case of SBS. Damages for the personal injury suffered are more difficult to quantify. The amount of damages will depend on the period of time for which the employee has suffered symptoms, the severity of the symptoms, whether they will continue and if so for how long.

VENUE: THE COURTS SYSTEM

The European Court of Justice

Article 1 6.4 of the Treaty of Rome provides that the Court of Justice shall ensure 'that in the interpretation and application of this Treaty the law is observed'. The Court of Justice can therefore be described as the final Court of Appeal in relation to the interpretation and implementation of all EC legislation. The European Court of Justice therefore:

- ensures compliance by member states with EC treaty obligations;
- gives rulings on EC law at the request of national courts;
- grants compensation for damage caused by EC institutions;
- acts as a Court of Appeal from decisions of national courts.

The English court system

Magistrates courts will deal with criminal proceedings for breach of duty under the Acts or regulations referred to. Serious cases will be dealt with by the Crown Court. County courts will deal with less complex civil claims for compensation. The High Court might deal with civil claims where the amount claimed exceeds £50 000; or where the issues involved are of general public interest; or where there are complex facts on legal issues involved; or where transfer may result in a speedier trial of the issues.

Thus most claims for compensation in relation to SBS would be likely to come within the county court jurisdiction on the grounds that the damages will be modest. However, if an employee were found to be unable to work in a particular building and suffered substantial loss of earnings his claim could conceivably be dealt with in the High Court. Equally, any test cases might well be transferred to the High Court if found to be particularly complex.

It is more frequently found in the 1990s that groups of people are entitled to claim compensation for a civil wrong, each person pursuing his or her individual claim. Such circumstances often arise when publicity is given to a new form of claim or where a trade union may wish to instigate proceedings on behalf of a large group of its members. In such cases the courts discourage the pursuit of a large number of separate actions. Lawyers at the forefront of such claims are encouraged to form a steering committee; often all claims are ordered to be referred to a particular High Court judge; a small number of cases will be pursued as test cases and the same judge will give directions and hear the eventual final trial. This procedure saves huge expense, is more specialized and can produce an earlier satisfactory outcome for claimants. If a large building were found to be affected by SBS it is possible that the class action procedure might be appropriate. Certainly the courts would discourage separate claims by each individual employed in a particular building.

CONCLUSION

Lawyers are continually widening the horizons of personal injury litigation. Many thousands of claims are brought annually in relation to personal injury suffered through work conditions. Industrial deafness, vibration white finger, repetitive strain injuries generally, pneumoconiosis, eye strain suffered by typists, mucous membrane disease are but a few of this category of claims brought in recent years. It remains to be seen if successful action can be taken to protect those suffering from SBS. This chapter has outlined the various difficulties faced by a potential plaintiff. It will not be long before such claims are tested in the courts.

REFERENCES

1. Report of the Committee on Safety and Health at Work, Cmnd 5034 1972.
2. *West Bromwich Building Society* v. *Townsend* (1983) 1 CR 257 DC.
3. Health and Safety at Work Act 1974.
4. *Pickering* v. *Liverpool Daily Post 1991*.
5. *Francovich* v. *Italian Republic* (1992) 1 RLR 84.

CHAPTER 9

Design for manageability

Adrian Leaman and Bill Bordass*

INTRODUCTION

Since the mid-1980s when the first studies of SBS were under way in the USA and UK, much has changed in the public perception of buildings. Health scares over Legionnaires' disease and international efforts to reduce energy consumption in buildings, among other factors, have helped to encourage research effort on the human, social and environmental aspects of buildings, rather than just on technical, constructional and design topics. At first, many of the findings from this new work were treated sceptically, especially by the building professions; but now its perspectives are fundamentally affecting thinking about building management and design. In this chapter, some of the pointers are followed up, and implications examined, especially for strategic thinking about buildings and their occupants.

Why do too many buildings deliver less than they promise? Barring major technical failure, a common answer is unmanageable complexity. Many buildings are prone to this fault, especially the newer ones which try to integrate a greater number of activities at higher intensities and spatial densities and with better amenities than in the past.

In striving for improvements, designers often underestimate or ignore:

- how systems – physical and human – can conflict with each other, thereby pulling performance levels down to lowest common denominators levels; and
- how uncertainty and inefficiency in systems' operation and use can readily develop through lack of attention to the detail of occupants' requirements.

Conflicts between systems and uncertainty in their use are symptoms of unmanageable complexity, a feature of modern buildings which arises

---

* A revised version of this chapter was published in Leaman, A. (ed.) (1995) *Buildings in the Age of Paradox*, University of York.

from the tendency, first, to require too much of the building, and then too much of its management. Often buildings are not designed with management and use in mind, and so can exhibit pathological characteristics – unnecessary overuse of fossil fuels, chronic illnesses of occupants such as dry eyes, hot, dusty and noisy spaces, absenteeism, productivity losses, uncontrollable indoor environments and low user morale. Many of these features are interrelated with the culture of the occupying organization, and it is often difficult to attribute direct causes. Chronic features also tend to reinforce each other, so that once standards slip the process becomes difficult and expensive to reverse.

These observations come from research on occupied buildings carried out in the United Kingdom over the past decade. In this chapter, we offer some pointers which may be incorporated into strategic thinking about building design and use, and especially some of the principles which should be introduced at the briefing stage of a project.

Most of the findings are based on projects with which Building Use Studies and William Bordass Associates have been directly involved: post-occupancy evaluations of new office buildings, studies of building services and energy consumption in offices, hotels, factories, retail, education and sports buildings, surveys of occupant ill health and occupant control behaviour, and the effectiveness of active and passive design features.

GENERAL FINDINGS

Figure 9.1 illustrates some of our general findings. There are two attributes on the diagram.

Systems in buildings may be considered as either physical (top half) or behavioural (bottom half). Treated as integral systems, including both physical and behavioural elements, most buildings are a mixture of tightly coupled and loosely coupled elements with interfaces between them [1]. Physical systems (such as the building structure, walls and enclosed spaces, windows and ventilation systems) tend to be tightly coupled, meaning that there is relatively little slack or give between them [2]. Behavioural systems are loosely coupled (meaning that certain parts express themselves according to their own logic or interests [2]).

Attributes may be context free (left-hand side) or context dependent (right-hand side). Context-free attributes are systems and principles that can be applied to buildings independently of their operation. These should include:

- features of buildings which should properly be relied upon to operate in the background, and be normally not noticed in everyday use;
- most technical systems;
- legislation governing building design and use.

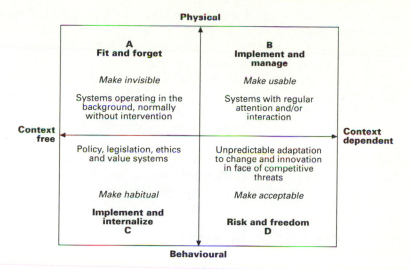

Fig. 9.1 Strategic design and management considerations.

Context-dependent attributes need to be tailored to suit the requirements of the occupants, and need regular attention or action.

In addition:

- the top left of the diagram represents characteristics which are predominately spatial, and normally created by designers who are usually outside the occupier's cultural system; and
- the right-hand side is the province of occupiers, users and managers who are usually much more preoccupied with time-dependent systems.

THE FOUR QUADRANTS

These two dimensions divide Fig. 9.1 into four quadrants, which we have named:

- Physical and context free: something which can be taken care of physically and does not alter with operational context can be seen as 'fit and forget'. The location of most buildings certainly is; so to a greater extent are passive features such as structural stability, fire compartmentation and insulation.
- Physical and context dependent: the demands on these aspects are forever changing and they do more than respond passively. They include equipment which needs reconfiguring, replacing and servicing; furniture which needs to be moved about; and engineering systems which

react to changing weather and occupancy. They need to be implemented and managed.

- Behavioural and context free: these are things which one can take for
  granted in (or at least reasonably expect from) people. They are
  ingrained in social structures and habits, ethics and value systems,
  and supported by government policies and rules. For things to go
  smoothly, however, what one wants should be implemented and inter-
  nalized, and it is usually much easier to go with, than against, the
  grain.
- Behavioural and context dependent: the unexpected happens; something goes wrong. All is going well until a telephone call changes
  everything! This is an area of risk, but freedom too.

AVOIDING UNNECESSARY DEMANDS

In general, the fewer demands a building makes on its occupants and its
management, the more likely it is to work as intended. However, no
building is infinitely durable and increasingly investors and occupiers
want buildings to be 'flexible'. Nevertheless, we can identify important
aspirations in all four quadrants of Fig. 9.1.

- Top left: make invisible. Ideally, these are things that one can fit and
  forget and which do their job invisibly, without intruding on the occupancy and use of the building.
- Top right: make usable. Things that need more changing around and
  looking after should be usable, and ideally by those most directly connected with them: it is better if you can move your own table and
  adjust your own thermostat and light. It is better if you can get at the
  item needing maintenance rather than having to disconnect other
  things and lift them out of the way.
- Bottom left: make habitual. Designers may expect occupants to behave
  in unfamiliar ways. Occasionally this may be necessary: but if so a
  strategy needs to be carefully worked out, discussed and agreed with,
  and implemented by management. However, if what you want people
  to do fits in with the way they normally do things, it makes life much
  easier. If it is intuitively obvious, better still.
- Bottom right. Make acceptable. Most hazards can be reduced to
  acceptable levels by a combination of physical, behavioural and managerial measures in the other three quadrants, plus risk management
  procedures. Few can be eliminated, at least at sensible cost (spending
  too much on reducing one kind of risk can easily divert funds from
  better and more cost-effective measures) and without unreasonable
  restrictions on freedom. Risks can also have a nasty habit of being
  shunted around: people in safer cars kill more pedestrians and cyclists
  [3]!

UNINTENDED CONSEQUENCES

Many problems with buildings seem to occur because people either put things in the wrong quadrant, or fail to appreciate that they belong in several quadrants. For example, to the occupant, an open plan office with air conditioning behind the ceiling (or under the floor) may appear to offer the ultimate in flexibility. However,

- The system will always have some intrinsic limitations, which always seem to surface sooner rather than later.
- All the equipment behind the scenes will need looking after by somebody. Has it been made usable for them?
- The individual occupants and groups in the space may find it more difficult, say, to alter their furniture or their temperature than transfer these activities to management; and if response is not rapid, they may become highly critical.

The result is that rather than 'fit and forget' there is quite a big task in routinely looking after the facilities which were intended to provide the flexibility. If this maintenance is not well done, the consequences can be serious. 'Fit and manage the consequences' (top left and right) might be a better phrase for it.

Perhaps it would be better to start off with something simpler, which makes less routine demands of management, even though it may require more substantial *ad hoc* interventions from time to time.

DESIGNER AND USER PERSPECTIVES

Although both designers and users usually try to create flexible buildings that respond well to changing requirements, they do so from different perspectives which are often incompatible. Designers tend to see buildings from the point of view of spatial constraints; users and occupiers from the perspective of time. The designer's perspective tends to be biased in favour of the left-hand side of Fig. 9.1: the user's from the right-hand side. Designers often stereotype or simplify user behaviour, or ignore it altogether [4]. Users often misunderstand or ignore the spatial and technological, cost and legislative constraints within which designers must operate.

As the authors have shown [5], designer and user perspectives can be complementary – especially when buildings are shallow in plan form and have simple heating and ventilation services – but tend to 'fight' each other as soon as they get bigger and more complex. The rapid growth of the facilities management profession in the USA, Europe and Australasia in the 1980s and 1990s has partly come about to deal with conflicts and inefficiencies created by large, complex buildings in which design and user issues have not been clearly enough resolved.

In the authors' experience:

• Too much attention is given to visible spatial features of buildings at the expense of less obvious time-dependent features.
• Many unintended consequences arise from trying to assign building system attributes to the wrong part of the diagram, or not recognizing the interactions between the various parts.

Many make the mistake that buildings can be designed and successfully run to standard procedures and performance specifications. This is rarely the case because of differences created by unique requirements of occupants and organizational cultures. Our experience is that building functions must be recognized for what they are and allocated appropriately; otherwise chronic, and occasionally catastrophic, problems will result.

Overreliance on technology, and burgeoning legislation (designed in part to deal with technological overkill) tends to want to push functions into the left-hand side of Fig. 9.1. The outcome is that buildings are becoming: harder to manage effectively, and surprisingly often less easy to change. At present, the vector of change seems to favour the top left of the diagram, with more standards and codes to be met and less scope for discretion in design and management. We may be seeking too many context-free physical solutions to problems which belong (at least partly) in other quadrants, and often turn up there whether we like it or not. By expecting too much of the building and too little of management, a self-reinforcing prophecy is created if the very process in fact places additional demands of a different kind on management and makes it more difficult to make appropriate and useful compromises.

There is a danger that this trend could be self perpetuating as designers, managers and legislators continue to seek technological solutions to what should more properly be considered human management problems.

The implication is that more attention should be given to understanding outcomes of human behaviour in real contexts, especially in respect of :

• risky, abnormal or dangerous circumstances;
• decisions made when individual actions are further constrained by group behaviour, including how individuals and groups respond to suboptimal internal environmental conditions;
• change, flexibility, adaptability and responsiveness of conditions to new situations;
• effects on behaviour and decision-making of changing work tasks;
• usability of control interfaces.

Each of these falls properly in the right-hand part of Fig. 9.1, and all have been relatively ignored in the recent past.

SOUGHT AFTER ATTRIBUTES

Table 9.1 summarizes attributes of buildings which studies have shown to be beneficial or sought after. These attributes could form the basis of a strategic brief for new or remodelled buildings. For the following sections, the supporting evidence will be found in the references.

Rapid response

Speed of response is a topic rarely covered in the building literature, although widely in management science [6]. The faster a building (meaning the whole building system, human as well as physical) can respond to requests for change from the occupants, the better people tend to like it and the more productive they say they are in it [7]. Speed of response applies in obvious ways such as the time taken by lifts to answer calls, or the time taken for a computer system to respond to a log-in request (four seconds is the tolerance threshold!) [8]. More emphasis is being placed on the speed with which furniture systems can be reconfigured, and possible cost savings by much more efficient relocation logistics.

Management procedures which react promptly to occupants' complaints also seem to be appreciated, even if the source problem cannot be entirely solved. Where quick response is the norm, whether through physical control systems such as adjustable blinds or manually adjustable thermostats, through building management support services, or a combination of both, occupant perceptions will usually be more positive and appreciative [9].

Table 9.1 The best buildings

1. Respond rapidly and positively to triggers of change at all spatial levels (individual, workgroup and department).
2. Have enough management resources to deal with adverse or unpredictable consequences of physical or behavioural complexity.
3. Are comfortable and safe for the occupants most of the time, but use the properties of 1. if they become uncomfortable or unsafe.
4. Optimize relationships between physical and human (managerial) systems at all lifecycle stages (such as briefing, design, commissioning, use).
5. Are economical of time in operation for all user types (individuals at their workplaces, workgroups and visitors).
6. Keep resource inputs to a necessary minimum, as well as minimizing undesirable effects which potentially infringe the rights of others.
7. Allow higher levels of functional integration to be retrofitted, if needed.
8. Do not introduce non-reversible failure pathways.

One reason why occupants appear to prefer openable windows in many situations is that they have fast response and intuitively obvious control, even though opening windows may not always deliver optimal or even reasonable conditions.

It is also important to have rapid response to failures within the technical systems. At present there are reasonably effective automatic systems to alert one to critical faults, for example a fire or a boiler lockout. Other faults (for example, lights failed) are quickly noticed by the occupants. Much less noticeable are chronic faults which affect efficiency but not service – or at least not very noticeably. Examples include:

- wasteful operation of heating and air conditioning systems, sometimes even running continuously;
- undetected malfunctions of energy-saving devices, such as heat recovery, free cooling and night ventilation systems.

Surveys often reveal that buildings designed to be energy efficient but which perform disappointingly suffer failures of this kind. For example, a review of case studies [10] found that differences in energy use depended more on the detailed design, commissioning, control, operation and management than on the technical features adopted. Human management was at least as important as technology in securing good energy performance, particularly in air conditioned buildings which had more potential for wastage.

Sufficient resources

Rapid response will usually be found in buildings where management has enough resources to deal with building-related problems both as and when they arise, and in advance. Good management will endeavour to set up self-reinforcing virtuous circles of causation which consistently 'deliver' quality and responsiveness. However, most buildings are the victims of vicious circles which can become increasingly expensive to halt or reverse and spiral into accelerating decline [11]. For example, vandalism encourages further decline unless an environment is cared for: with immediate repainting or repairs, when the process can often be stalled [12].

As often as not, the true costs of running buildings are underestimated or ignored altogether by designers and senior management, forcing many buildings into vicious circles from move-in day. Building budgets are soft targets for cutbacks, partly because line managers do not have convincing data with which to defend themselves against attack from above. But much can be done in good briefing and design to reduce the management task by making things less complex and more self-managing. As a rule of thumb, based on data from the Building Services Research and Information Association (BSRIA) and from Bernard

Williams Associates [13, 14], the annual spend on building services maintenance should be about the same as that for energy. This does not guarantee success, of course; but if the figures differ widely something may be wrong, especially if the energy spend is high and the maintenance spend low.

The general experience of Building Use Studies is that maintenance of buildings leaves a great deal to be desired, either from knock-on effects of chronic long-term underfunding (as in many British schools for instance) or through bad maintenance habits and practices, including the appointment and supervision of outside contractors. Early work on SBS in UK offices led many, including the authors, to think that SBS was primarily a design problem (with the main explanatory variables being physical features such as type of ventilation system or depth of space). As understanding grew, it became clearer that management, and maintenance, variables were more important than first thought [15, 16].

Designers and clients seeking flexibility, or energy efficiency, may unwittingly add to the management resource requirement and hence sow the seeds of failure. For example, one report [10] noted that 'complex energy systems may not be operated as the designers intended, and saved heating and cooling energy may turn up instead as parasitic losses from pumps, fans and unforeseen control problems. The greatest savings nationally are likely to come from simple applications of available technology in a manner which integrates architectural, engineering and user requirements, and provides control and management systems to suit.'

Alleviating discomfort

One of the best kept secrets of work on thermal comfort in buildings is that alleviating discomfort is just as important for occupants' satisfaction as providing comfortable conditions in the first place [17, 18]. Occupant dissatisfaction with the indoor environment is directly related to occupants' perceived productivity [19] – the link between dissatisfied staff and better productivity. On this basis, it may be better to give building occupants more capability to fine tune their environment than to rely too much on fully automated systems which in theory can deliver a better environment but may not be perceived as doing so.

Designers often assume that comfort can be achieved solely by systems designed to 'keep the measured variables within the required tolerances' and leave out the other features. The best buildings for comfort and energy efficiency require all four features shown in each of the quadrants of Fig. 9.2. They need automatic control (top half of diagram) plus manual control (bottom) and if possible should anticipate likely change (right half) and not just operate in response mode (left half). However, gratuitously adding more controls may introduce conflicts between different subsystems and increase complexity beyond manageable bounds. For

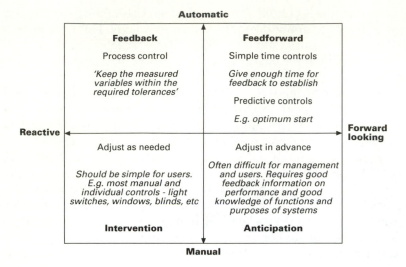

Fig. 9.2 Control strategies for building services.

example, open plan offices trade off the personal controllability normally found in cellular spaces for greater interpersonal communication in the open areas (at least in theory, for many in surveys report it as an annoying distraction!). Productivity gains from better communication may not outweigh the productivity losses caused by more distracting, less controllable and usually hotter environments.

User control is important because people are often better than preprogrammed systems at dealing with unusual or unpredictable situations. The number of unusual situations is also likely to increase as space use is intensified. Like modern airline pilots who normally fly under autopilot but take control in difficult, unusual or emergency circumstances, building users need the capacity to make adjustments; their tolerance of conditions also increases as perceived control rises. For example, users seem to accept 'poorer' conditions in naturally ventilated than in air conditioned buildings [20].

These considerations also apply in the arena of safety and health, and especially in the rapidly growing subject of risk assessment. Table 9.2, adapted from [21] briefly illustrates some of the considerations. See also [2].

Unfortunately, some engineering and energy-saving systems may create rather than alleviate discomfort. As a general rule it appears that:

- manual systems should operate perceptibly and give immediate response, if not by performing the intended function then at least by giving a click or lighting an indicator;

Table 9.2 Risk estimation considerations

Failure to consider the ways in which human errors can affect
technological systems
*Example: Obscure and difficult to operate Building Management Systems
resulting in energy wastage and discomfort.*

Overconfidence in current scientific knowledge
*Example: Failure to take unproven scientific evidence seriously or develop
precautionary strategies (e.g. global warming).*

Failure to appreciate how technological systems function as a whole
*Example: Overlooking importance of control interfaces in buildings,
especially manual controls.*

Slowness in detecting chronic, cumulative effects
Example: Building-related sickness.

Failure to anticipate human response to safety measures
*Example: Windsor Castle fire where emergency telephones were not seen by
those wishing to raise the alarm.*

Failure to anticipate common-mode failures, which simultaneously
afflict systems which are designed to be independent
*Example: Failure of innocuous window components like friction hinges in
naturally ventilated offices simultaneously affecting noise, ventilation
and heating performance.*

Source: Adapted from [21]

- automatic systems should operate imperceptibly: if not, whatever they
  do is sure to be wrong for some occupants.

Automatic control of lighting and blinds are common offenders here [22]:
the blinds close either just as you are enjoying the sun or long after you
have become fed up with it; the lights come on when you enter the room
whether you think you need them or not; and other people's lights flash-
ing annoy you. Automatically controlled windows in new 'green' buildings may create similar problems. Individual user overrides of such
systems are not costly luxuries, they are essential.

Optimize relationships between physical and human systems

Although buildings and their occupying organizations are recognizably
complex systems, with many levels of interaction and feedback between
subsystems, many are designed, built and occupied as if they were independent systems with simple causality. It is commonplace to hear designers plead for their specialism (lighting, security, furniture and so on) to

receive priority in the design process. This way they can avoid or minimize constraints deliberately or unwittingly imposed by others, and perhaps pass on some of their own for good measure!

True integration, with attention to detail and avoidance of unnecessary
conflicts, comes through a well-developed briefing process which does
not compromise the specialist designers' role. Later in the building's life,
the brief should become the yardstick for post-occupancy surveys which
objectively test whether it was met. This information may then be fed
into new building briefs, closing the quality improvement loop. The now
extensive literature on 'total quality' offers many suggestions for building managers. For instance, techniques used in small-scale product
development seem particularly appropriate to use at the larger building-
system level [23].

For building and environmental services, it is important that the point
of control is as close as possible to the appropriate point of need.
Anything else will require access to management resources: which is at
best wasteful, and usually means that an undesirable state becomes the
default state because it is the most convenient.

Economy of time

Buildings operate over time as well as in space but far more attention has
been given to performance in relation to spatial variables (like depth,
height, shape and form) than over time. As a result, space and time systems are often poorly integrated and physical solutions proposed where
operational approaches may have been better, and vice versa. In future,
much more thought will be given to the way buildings work dynamic-
ally, especially to overcoming utilization inefficiencies. Understanding
time properly involves not just considering gluts and famines of occupancy, but also factors such as how habits, attitudes and behaviour influence the way systems work.

The best buildings keep to a necessary minimum time wasted by occupants moving about. This point is closely related to response times – the
faster the need is met, the better. This applies not just to more obvious
facilities such as the location of meeting-rooms or toilets, but also activities such as photocopying, with major inefficiencies in queuing, machine
downtime and travel time to the machine location.

Buildings too often default in performance to undesirable states which
are extremely hard to alter. For example, many run with all their lights on
all day because the first person who arrives in the morning in the half-
light of dawn will switch all the lights on (at the gang switch near the
door). Perhaps they have no option, perhaps the switching is incompre-
hensible, or perhaps they just want to 'cheer the place up'. As successive
people arrive, it becomes harder and harder to switch any lights off
because of the difficulty of agreeing among everyone that this should

happen. The building will thus tend to run 'lights on' by default, whatever the daylight conditions outside. The combination of habit, poor control design, and the difficulty of making small-scale 'trivial' decisions in groups leads to unnecessary inefficiency and suboptimal working environments. Here, lack of integration between spatial factors and time factors (location of light switches, times of arrival) leads to buildings running 'just in case' – that is, inefficiently and insensitively to true demand. Automatic daylight linked controls are not the complete answer to this problem, as discussed above. Human and automatic systems need to be sensitively combined.

Economy of time in fact unites all the features 1–5 in Table 9.1. A simple rule is to make 'the bad difficult and the good easy', which means comprehensible devices correctly located, easily operated, and configured to give rapid response and avoid unnecessary waste.

Sufficient resources

The best buildings match demand and supply and keep 'just in case' running to a necessary minimum. Buildings which work best for human comfort and satisfaction also tend to be energy efficient [24, 20] probably because a good match of demand and supply is achieved through careful performance monitoring, attention to users' complaints and relatively rapid feedback loops and well-defined diagnostics. This is helped along by robust, well-designed, user-friendly systems. Effective cleaning and maintenance and efficient energy management all involve active monitoring of systems' performance. The cleaning or the energy saving may not be the most important part of these activities, but the monitoring and the culture which causes it all to happen [25].

Buildings are undergoing a demand-side revolution of which the rapid growth of the facilities management professions is an important part. Emphasis on systematic building evaluation techniques is increasing, in an attempt to give potential occupiers a clearer understanding of strengths and weaknesses in advance of committing themselves to leases or purchase.

Through wider understanding of building performance – through investment, costs in use, technical features and human factors – clients are much more aware of the right questions to ask their design teams. Faced with an informed client, and far more focus on problem definition, designers must respond with better predictions of what their buildings will deliver. Architects and engineers now have less influence over briefs and the basic strategic agendas for buildings. This is not necessarily a bad thing, because more attention to needs and requirements provides designers with better problem definition in the building brief, potentially enabling them to give a better response.

Higher levels of integration

The best buildings allow more functions to coexist, and are tolerant of higher levels of functionality. This is most apparent when buildings are altered to suit new requirements. Almost invariably, the altered space will be more densely occupied and accommodate a wider range of activities, for example, in higher education buildings which change uses from daytime to evening and from term-time to vacation and converting offices from cellular to open plan. The best buildings are able to accommodate higher densities as well as more functions operating simultaneously. However, there is a discernable trend both towards greater space intensification and increased obsolescence.

The desire for (or promise of) 'flexibility' often leads to solutions which are reliant on energy-dependent technologies such as air conditioning. However, in practice this flexibility may not be as great as was initially hoped, witness all the materials, many often nearly new, which end up on the skip when an office is fitted out. An alternative route may sometimes be to provide a simpler, but potentially adaptable, building, but one which is easily altered as needs change. If properly thought through, this can reduce both initial and in-use costs. 'Mixed mode' services concepts allow natural ventilation and mechanical systems to work together [26].

Minimize failure pathways

Few buildings fail catastrophically in a technical sense. Many more fail economically, functionally, aesthetically or socially and exhibit chronic failures of one kind or another which often last for the lifetime of the building because there are no reasonable means of correcting the fault once it is there.

With the benefit of hindsight, some of these once latent faults seem blatantly obvious, but they can be hard to detect unless thorough briefing and design management disciplines are in place, plus appropriate testing of solutions. With the development of risk analysis techniques, which help prevent accidents in complex and dangerous systems like nuclear power plants [2, 27] one can now begin to target areas of most risk and put prevention strategies in place early in the design process. For example, in a naturally ventilated building, the window is one of the most crucial building elements, so it is imperative that the window elements should operate reasonably effectively and in sympathy with associated systems, or failure in apparently 'trivial' components can be excessively costly in the long term.

CONCLUSION: DESIGN FOR MANAGEABILITY

Most of the pointers introduced here lead to the single conclusion: design for manageability. For manageability's sake:

- The fewer demands a building makes on management services, the better.
- Passive is better than active. Make sure that things which are designed to operate in the background properly do so.
- Things which need changing or looking after should be usable, preferably by those who are most directly concerned with them. Responses should be rapid and understandable.
- Simple is better than complex, but when complexity is necessary package and isolate it wherever possible, and provide simple interfaces.
- Cater where possible for people's preference ranges rather than the average or norm. Try to foresee risky situations and how people may compensate.
- Identify potential failure paths and try to avoid them; if not, monitor appropriate indicators to help identify, and deal with, incipient problems.
- Beware of unsubstantiated promises of 'flexibility' which may bring unforeseen management costs. Recognize that all situations are subject to constraints, which will show themselves sooner or later.
- Try to assess risk cost effectively, so that resources are realistically spent on avoiding the costliest and most risky events.
- Remember that designers are not users, although they often think they are!

REFERENCES

1. Simon, H. A. (1981) *The Sciences of the Artificial*, 2nd edn, MIT Press, Cambridge, MA.
2. Perrow, Charles (1984) *Normal Accidents: Living with High Risk Technologies*, Basic Books, New York, p. 91.
3. Adams, John (1987) *Risk and Freedom*, Newnes, London.
4. Norman, Donald A. (1988) *The Psychology of Everyday Things*, Basic Books, New York.
5. Leaman, A.J. and Bordass, W.T. (1995) Comfort and complexity: unmanageable bedfellows? *Workplace Comfort Forum*, RIBA, London.
6. Meredith J.R. (1992) *The Management of Operations: a Conceptual Emphasis*, 4th edn, John Wiley, London.
7. Bordass, W.T., Bromley, A.K.R. and Leaman, A.J. (1995) *Comfort, Control and Energy Efficiency in Offices*, BRE Information Paper, IP3/95.
8. Neilsen, J. (1993) *Usability Engineering*, Academic Press, London. (This book makes several references to the importance of response times in the context of software interfaces. For example, a computer system should pay more attention to a user's new actions, giving them higher priority than finishing old tasks.)

9. Bordass, W.T., Leaman, A.J. and Willis, S.T.P. (1994) Control strategies for building services: the role of the user. *Chartered Institute of Building Conference on Buildings and the Environment*, Building Research Establishment, UK.
10. Energy Efficiency Best Practice Programme (1994) *Technical Review of Office Case Studies and Related Information*, General Information Report GIR 15 (Brecsu/EEO).
11. Hampden-Turner, Charles (1990) *Corporate Culture: from Vicious to Virtuous Circles*, Economist Books, Hutchinson, London.
12. CIRIA (Construction Industry Research and Information Association) (1994) *Dealing with Vandalism*, CIRIA Special Publication 91, London.
13. Smith, M.H. (1991) *Maintenance and Utility Costs: Results of a Survey*, BSRIA technical memorandum 3/91 (quoted in Armstrong, J., The management of maintenance. *Building Services and Environmental Engineer*, May 1994).
14. Williams, B. (1993) The economics of environmental services. *Facilities*, **11**(11), pp. 13–23.
15. Wilson, S., O'Sullivan, P., Jones, P. and Hedge, A. (1987) *Sick Building Syndrome and Environmental Conditions*, Building Use Studies, London.
16. Leaman, A.J. and Tong, D. (1994) The indoor environment: strategies and tactics for managers, in *CIOB Handbook of Facilities Management* (ed. A. Spedding), Longman. London.
17. O'Sullivan, P. (undated) Criteria for the thermal control of buildings: people, Welsh School of Architecture.
18. Humphreys, M.A. and Nichol, J.F. (1970) An investigation into the thermal comfort of office workers. *JIHVE*, **38**, pp. 181–9.
19. Leaman, A.J. (1994) Dissatisfaction and office productivity. *The Facility Management Association of Australia*, Sydney.
20. Bordass, W.T. and Leaman, A.J. (1993) Control strategies for building services: advanced systems of passive and active climatisation, Institut Catala d'Energia (ICAEN) as part of Thermie programme (available from authors on fax 44 01904 611338 if source difficult to trace).
21. Bromley, A.K.R., Bordass, W.T. and Leaman, A.J. (1993) Are you in control? *Building Services: The CIBSE Journal*, April.
22. Fischhoff, B. (1989) Risk: a guide to controversy. Appendix C of *Improving Risk Communication*, National Research Council, Washington, DC.
23. Bordass, W., Leaman, A., Heasman, T. and Slater, A. (1994) Daylight in open-plan offices: the opportunities and the fantasies. *Proceedings of CIBSE National Lighting Conference*, Cambridge, UK, pp. 251–9.
24. Ernst and Young Quality Improvement Consultancy Group (1994) *Total Quality: A Manager's Guide for the 1990s*, Kogan Page, London.
25. Leaman, A.J. and Bordass, W.T. (1994) The dirt devils: cleaning and the culture of responsiveness. *Safety and Health Practitioner*, February.
26. Bordass, W.T., Entwistle, M.J. and Willis, S.T.P. (1994) Naturally-ventilated and mixed-mode office buildings: opportunities and pitfalls. *CIBSE National Conference Proceedings*, Brighton, UK, pp. 26–30.
27. Reason, J. (1990) *Human Error*, Cambridge University Press, Cambridge.

CHAPTER 10

Assessment and rectification

Jack Rostron

INTRODUCTION

It is important that buildings provide a healthy, safe and comfortable environment for occupants. Considerable attention has recently been given to the high incidence of sickness amongst people who work in modern office buildings. SBS is not only of obvious concern to the sufferer, but has commercial implications, in terms of increased absenteeism, reduced productivity, increased staff turnover, low morale, etc. This chapter will help owners, developers, facilities managers, architects, surveyors and other professional advisers in assessing existing and potential office buildings. It will also be of assistance to students in understanding this relatively new but important phenomenon.

The first section of the chapter explains SBS and its possible causes. It highlights and evaluates factors which should be taken into account in determining if it is likely to exist or how it can best be eliminated.

The second section offers a checklist to assess those factors which need careful attention in order to prevent the existence of SBS, i.e. ventilation, humidity, heating, lighting, contaminants, furnishings and colour scheme, maintenance, cleaning, use of the building, building management and noise. The checklist is intended as an indicative guide. Each question in each section should be recorded. Subtotals should be recorded for each factor and also be recorded on the summary sheet. Where a section provides a high proportion of negative answers, considerable remedial work or alteration will be necessary. If the grand total shows a very high proportion of negative answers, relocation may be the preferred option.

SBS EXPLAINED

SBS by its nature is difficult to define. It is generally considered to be a group of symptoms which people experience specifically at work, the typical symptoms being:

- lethargy;
- loss of concentration;
- nausea and dizziness;
- headache;
- hoarseness, wheezing and itching;
- skin rash;
- eye and nose irritation.

While the population as a whole generally exhibit these symptoms, with SBS, certain patterns evolve:

- The symptoms disappear or decline away from work.
- They are more prevalent in clerical staff.
- They occur more in public buildings.
- They are most common in office buildings with air conditioning.
- People with most symptoms have little individual control over their environment.

SBS should not be confused with Legionnaires' disease.

CAUSES

SBS is generally considered to result from one or more of the following factors:

- uncomfortable working environment due to poor lighting [1, 2], high temperatures and inadequate air movements/stuffiness [3].

- low relative humidity [4, 5];
- odours [6];
- air-borne dusts and fibres [7];
- chemical pollutants [8].

Ventilation

The ventilation system is often regarded as the most significant factor in affecting buildings which are sealed and have mechanical ventilation or air conditioning. The assumption is that lack of fresh air is the major cause of SBS.

Fresh air is required for various reasons, the main ones being to supply air for respiration and to dilute CO_2, odours, cigarette smoke and other contaminants. Ventilation, although not necessarily fresh air, may also be required to maintain personal comfort, i.e. for the control of air temperature.

Various standards exist for ventilation and fresh air supply rates to offices. They range from 5 litres per second per person in general offices up to 25 litres per second per person for personal offices or boardrooms where smoking is heavy.

The impetus to seal buildings and increase the control over the environment is usually motivated either by necessity for open plan deep offices which are difficult to ventilate naturally or by a desire to save energy (and money). The practice of tight control over the indoor environment poses problems if the ventilation or air conditioning system is in any way imperfect.

Mechanical ventilation of buildings is less satisfactory than natural ventilation because:

- Mechanical ventilation and air conditioning allow more precise overall environmental control but little personal or local control.
- The air supply into mechanical ventilation systems can often be varied during operation, in order to increase the proportion of air that is recirculated, and to reduce the quantity of fresh air drawn in from outside.
- Mechanical ventilation and air conditioning systems have components that are susceptible to failure and to poor design or installation.
- Recirculating ventilation and air conditioning systems can harbour organic growth and may distribute pollution from one area throughout the building.

Inadequate fresh air is probably a contributory factor rather than a sole cause of SBS.

Humidity

It is necessary to be able to control humidity in the workplace for a variety of reasons. Very high humidity can cause discomfort, especially at elevated temperatures and may result in excessive condensation. Low humidity causes drying of the mucous membranes resulting in eye, nose and respiratory discomfort. For most offices a relative humidity of 40–60% is appropriate and will prevent the build up of static electricity.

Humidification is important for comfort and health; but if the humidifier is allowed to become contaminated with micro-organisms and distribute contaminated water from humidifiers or from air washers, it

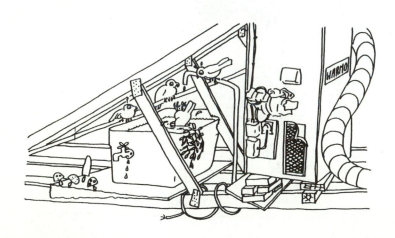

can cause various illnesses, such as 'Monday sickness' or 'humidifier fever'.

Several types of equipment exist for controlling humidity. In assessing the options, the primary object should be to prevent dispersion of heavily contaminated water droplets from humidifiers or air washers. The systems which are least likely to contribute to SBS in descending order, are:

- steam humidification;
- hot water evaporators;
- cold water evaporators;
- spinning disk and spray humidifiers;
- air washers.

The humidifier water supply should be clean and free from contamination. Water supplied directly from the mains greatly reduces the risk of contamination. The humidifier and storage tanks or reservoir should be regularly and thoroughly cleaned. This is particularly important if the system is of a spray or atomizing type.

Environmental comfort

Various standards have been set for the comfort of building occupants, the most widely accepted being the international standards (ISO 7730-1984). Recommended comfort requirements are:

- operative temperature 20–24°C (23°C is normally accepted for the UK);
- vertical air temperature at head and ankle height should show less than 3°C variation;
- floor surface temperature 19–26°C (29°C with floor heating systems);
- mean air velocity less than 0.15 m/s.

Dissatisfaction with the thermal environment is a greater problem in large air conditioned buildings than in small and naturally ventilated buildings. In a building with opening windows and radiators the occupants are able to vary the thermal environment to some extent. If the air conditioning or heating system in a large 'tight' building fails to control the thermal environment, there is often little that the occupants can do to improve conditions.

A sensation of 'stuffiness' may play a part in promoting SBS, indicating dissatisfaction with the working environment.

Visual environment

Potential problems in the visual environment are inadequate illumination, uniform or dull lighting, discomfort glare, flicker from luminaires and tinted windows which reduces the amount of daylight. These cause eye strain and headaches and are a major contributor to SBS.

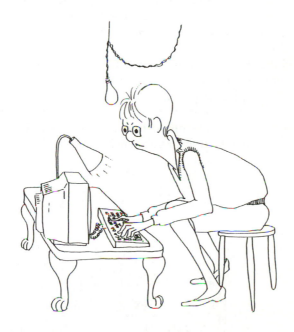

It is generally accepted that there is a link between the level of workers' satisfaction and their perceived ability to control the environment. Perhaps one of the most effective and economic solutions, especially in deep open plan offices, is the provision of task lighting.

The use of uplighting in open plan offices has greatly added to comfort levels. Similarly, the use of high frequency lighting is considered to greatly add to office worker productivity. The reduction of ultraviolet light by the installation of appropriate fittings is considered to reduce the symptoms of SBS by reducing indoor chemical pollution.

Contaminants

The potential range of contaminants in offices is enormous. The main sources of air-borne contaminants are:

- Building occupants: pollutants released by occupants of the building include CO_2, water vapour and micro-organisms and matter. Smoking is a considerable source of air-borne pollution.
- Building fabric and furnishings: the main sources of pollution are from releases (or 'offgassing') from the fabric and furnishings of the building; dust and fibres from carpets, and furnishings; solvent vapours and organics from various sources, including adhesives used in furniture and for sticking carpets, floor tiles, etc. Formaldehyde, especially from urea formaldehyde insulation and certain types of board, is an irritant, and may therefore cause some symptoms similar to those of SBS.
- Office machinery: photocopiers have been suggested as a cause of building sickness, and pollutants such as ozone can collect in very poorly ventilated photocopying rooms.
- Ventilation and air conditioning systems: ventilation and air conditioning systems can transmit air-borne disease including 'humidifier fever', and various infections. Even where air conditioning systems do

not contain humidifiers, items of plant can act as breeding sites for organic growth. This is true of items such as cooling coils where condensed water can collect, and these have been shown to release micro-organisms into the airstream.

Management and maintenance

- Efficient planning, particularly with the organization of office space and storage means less clutter and overcrowding. Untidy piles of papers and books not only create dust, but also collect dust as these areas are not easily cleaned.
- Management should be sensitive and people orientated, as this will promote goodwill and higher levels of satisfaction.
- Proper maintenance and regular cleaning of mechanical plant and ductwork are essential.
- The cleaning regime for soft furnishings, carpets and curtains should be carefully considered. Agents used should be chosen to eliminate potential sources of SBS and not inadvertently add to it.
- Files should be vacuum cleaned in order to remove paper and other dust as thoroughly as possible.
- Vacuum cleaners generally should be fitted with high efficiency final filters.
- Cool shampooing of carpets, chairs and other fabrics should be undertaken periodically.
- If symptoms persist steam cleaning should be considered.

Noise

Noise in itself has not generally been considered to be a main cause of SBS. It is clear, however, that both office workers' productivity and comfort levels can be affected by a poor acoustic environment. Most noise sources from both fixed plant and machinery and office equipment, can normally be silenced by appropriate physical measures.

In open plan offices, the maintenance of conversational privacy is important, and can often be achieved by the positioning of appropriate screens. The need for privacy suggests that cellular offices or several groupings of up to five workers in open plan offices helps reduce the symptoms of SBS.

CHECKLIST

This checklist will help you to determine if an office is a sick building. To determine if the building is sick, each question in each section should be recorded. This should be undertaken for the building as a whole, and depending on the circumstances, for each office in the building.

Then count up the number of 'yes' and 'no' answers and put the number in the subtotal box. When the grand total in the summary sheet shows a high proportion of 'no answers', an alternative building may be the preferred course of action.

However, the cost should also be borne in mind here, because even a lot of 'no answers' could mean that it is still worthwhile undertaking necessary remedial actions or changing the design.

Ventilation

1. Has the building been designed to Chartered Institution of Building Services Engineers (CIBSE) or American Society of Heating, Refrigeration and Air Conditioning (ASHRAE) recommended air change rates?   Yes   No
2. Are these rates achieved in practice?   Yes   No
3. Is the percentage of fresh outside air used in the system above 10%?   Yes   No
4. Are the air intake vents sited away from sources of contamination?   Yes   No
5. Is the prevailing wind direction likely to disperse potential sources of pollution?   Yes   No
6. Are the specified air filters for the plant used?   Yes   No
7. Are the filters fitted correctly?   Yes   No
8. Are there any openable windows to allow staff to ventilate space as required?   Yes   No
9. Have all air diffusers been connected to the ducting plant?   Yes   No
10. Are there any diffusers blocked by furniture?   Yes   No
11. Is there at least one inlet and one extract vent in each room?   Yes   No
12. Do air diffusers give correct air circulation?   Yes   No
13. Have spaces been tested for correct air change rates and dead spots?   Yes   No
14. Are exhaust luminaires used?   Yes   No
15. In the event of a plant breakdown is there an alternative source of fresh air provision?   Yes   No
16. Cold draughts do not exist.   Yes   No

Subtotal   ___   ___

Humidity

1. Is the relative humidity level maintained at between 40% and 60%?   Yes   No
2. Is a steam humidification system used?   Yes   No
3. Is it purged at least once per day?   Yes   No
4. Is the humidification system free from organic growth?   Yes   No
5. Is the humidification system fully serviceable   Yes   No
6. There is no carry over of water spray past the humidifier station?   Yes   No
7. Is a biocide or any chemical treatment used in the system?   Yes   No

8.	Is a weekly inspection of the system carried out?	Yes	No
9.	Are dehumidification coils correctly operated?	Yes	No
10.	Is there an air break on the condensate drain?	Yes	No
11.	Are there no static electricity problems in the space?	Yes	No

Subtotal   ____   ____

Heating

1.	Is the space temperature greater than 22°C in the heating season?	Yes	No
2.	Is the space temperature greater than 23°C in the summer?	Yes	No
3.	Are the temperature variations less than 3°C across the working space?	Yes	No
4.	Is the building zoned?	Yes	No
5.	Are there any elements in the space that will affect radiation asymmetry?	Yes	No
6.	Does the heating or air conditioning system have a terminal reheat facility?	Yes	No
7.	Do all the thermostats in the space function correctly?	Yes	No
8.	Have the occupants individual control over the heating/cooling at their workstation?	Yes	No
9.	Is the building a traditional heavyweight shell?	Yes	No
10.	Is the predominant orientation of windows other than south facing?	Yes	No

Subtotal   ____   ____

Lighting

1.	Is low frequency tubular fluorescent lighting avoided?	Yes	No
2.	Does the system operate on high frequency?	Yes	No
3.	Is an uplighter system used?	Yes	No
4.	Are specific luminaires used to alleviate screen glare on VDUs?	Yes	No
5.	Is task lighting available?	Yes	No
6.	Are CIBSE lighting levels achieved?	Yes	No
7.	Is glare avoided in the space?	Yes	No
8.	Are window shades available?	Yes	No
9.	Is solar reflective glass avoided?	Yes	No
10.	Are natural daylighting levels achieved?	Yes	No

11.	Are special shading provisions made on south facing elevations?	Yes	No
12.	Are there any problems of light contrast in the space?	Yes	No
	Subtotal	___	___

Contaminants

1.	There has been no refurbishment in the last year.	Yes	No
2.	New furniture has not been installed in the last year.	Yes	No
3.	Has the use of volatile organic compounds been avoided?	Yes	No
4.	Are photocopiers or printers housed in sealed rooms with their own extractor systems?	Yes	No
5.	Does the photocopier or printer exhaust system vent directly to atmosphere?	Yes	No
6.	Has the building been 'baked'?	Yes	No
7.	Is the building constructed on uncontaminated land?	Yes	No
8.	Has the existence of ureaformaldehyde insulation been investigated?	Yes	No
9.	Has the existence of asbestos been investigated?	Yes	No
	Subtotal	___	___

Furnishing and colour scheme

1.	Are plants and small trees located in the office space?	Yes	No
2.	Are furnishings, carpets and wall finishes colour coordinated?	Yes	No
3.	Has office furniture been ergonomically designed?	Yes	No
4.	Have staff been consulted on furnishings?	Yes	No
5.	Have furniture diffusers been considered?	Yes	No
6.	Have furnishings been assessed for fibre loss?	Yes	No
7.	Is there a high proportion of open shelving?	Yes	No
8.	Has specialist furniture for VDUs and computerware been considered?	Yes	No
	Subtotal	___	___

Maintenance

1.	Are air filters maintained as per plant manufacturer's specification?	Yes	No

2.	Are maintenance chemicals used in correct proportions?	Yes	No
3.	Are manufacturers plant maintenance schedules adhered to?	Yes	No
4.	Is there a planned maintenance system in operation?	Yes	No
5.	Are air flow rates at diffusers/vents as per design and commissioning specifications?	Yes	No
6.	Are regular inspections of plant above suspended ceilings or below modular floors carried out?	Yes	No
7.	Is the heating system regularly descaled and flushed?	Yes	No
8.	Are ceilings and walls regularly decorated?	Yes	No
9.	Are all condensate drains regularly checked and flushed?	Yes	No
10.	Is the lighting system regularly maintained?	Yes	No
	Subtotal	___	___

Cleaning

1.	Is the building fabric regularly cleaned, including exterior windows?	Yes	No
2.	Are internal surfaces including carpets, floors and furniture regularly cleaned?	Yes	No
3.	Does regular damp dusting take place on all hard surfaces?	Yes	No
4.	Are cleaning fluids and chemicals used correctly to manufacturer's specification?	Yes	No
5.	Is the cleaning plant used as per manufacturer's specification?	Yes	No
6.	Are air vents/diffusers regularly cleaned?	Yes	No
7.	Are luminaires regularly cleaned?	Yes	No
8.	Are air filters cleaned as per manufacturer's specification?	Yes	No
9.	Are ventilation ducts inspected and cleaned as necessary?	Yes	No
10.	Are heating/cooling coils regularly cleaned?	Yes	No
11.	Are the insides of filing cabinets regularly vacuumed?	Yes	No
12.	Are soft furnishings occasionally cool shampooed or steam cleaned?	Yes	No
	Subtotal	___	___

Use of building

1.	Is the building occupied by a private organization?	Yes	No
2.	Is the work principally of a managerial or technical nature?	Yes	No
3.	Is the building used as per the architect's design brief?	Yes	No
4.	Is the office layout cellular?	Yes	No
5.	Is the original occupancy level of the building achieved?	Yes	No
6.	Has additional electrical equipment in use in the space been taken into account with regard to the plant cooling load?	Yes	No
7.	Is dust and pollution from building alterations avoided?	Yes	No
8.	Does partitioning take into account the heating and ventilating system?	Yes	No
	Subtotal	___	___

Building management

1.	Is a computerized building management system in place?	Yes	No
2.	Is a remote system avoided?	Yes	No
3.	Is temperature and humidity checked by maintenance personnel?	Yes	No
4.	Do staff have a complaints procedure if they feel the working environment is unsatisfactory?	Yes	No
	Subtotal	___	___

Noise

1.	Are CIBSE noise reduction levels achieved in all spaces?	Yes	No
2.	Is the building designed with regard to acoustic problems?	Yes	No
3.	Are plant rooms constructed to achieve correct noise reduction levels?	Yes	No
4.	Are silencers fitted correctly to supply and extractor ducts?	Yes	No
5.	Is noisy machinery isolated?	Yes	No
6.	Are crosstalk attenuators fitted?	Yes	No

7. There are no sources of vibration within the
    plant rooms?   Yes   No
                                                          ——   ——

                                            Subtotal   ——   ——

SUMMARY SHEET   YES   NO

Ventilation   ——   ——

Humidity   ——   ——

Heating   ——   ——

Lighting   ——   ——

Contaminants   ——   ——

Furnishings and colour scheme   ——   ——

Maintenance   ——   ——

Cleaning   ——   ——

Use of building   ——   ——

Building management   ——   ——

Noise   ——   ——

                                     Grand total   ——   ——

REFERENCES

1. Wilson, S. and Hedge, H. (1987) *The Office Environment Survey: A Study of Building Sickness*, Building Use Studies, London.
2. Robertson, A. and Burge, S. (1986) Building sickness: all in the mind, *Occupational Health*, March.
3. Berglund, B., Berglund, U. and Lindvale, T. *et al.* (1984) Characterisation of indoor air quality and sick buildings, *ASHRAE Transactions*, **90**(1B), pp. 1045–55.
4. Gilperin, A. (1973) Humidification and upper respiratory infection incidence, *Heating/Piping/Air Conditioning*, March.
5. Green, G.H. (1984) The health implications of the level of indoor humidity, *Indoor Air*, **1**, Swedish Council for Building Research.
6. Guidotto, T.L. *et al.* (1987) Epidemiological features that may distinguish between building associated illness outbreaks due to chemical exposure of psychogenic origin, *Journal of Occupational Medicine*, **29**(2), pp. 148–50.
7. Reisenberg, D.E. and Arehard-Treichel, J. (1986) Sick building syndrome plagues workers dwellings, *Journal of the American Medical Association*, **225**, p. 3063.
8. Hicks, J.B. (1984) Tight buildings syndrome: when work makes you sick, *Occupational Health and Safety*, January.

Appendix A:
Useful addresses

The Biolab Medical Unit
The Stone House
9 Weymouth Street
London
W1N 3FF
Indoor air quality analysts
Tel. 0171–636–5959/5905

British Medical Association
BMA House
Tavistock Square
London
WC1H 9JP
Tel. 0171–388–8296

British Occupational Health Research Foundation
9 Millbank
London
SW1P 3FT
Tel. 0171–798–5869

Building Performance Services Ltd
Grosvenor House
141–143 Drury Lane
London
WC2B 5TS
Architects, engineers and surveyors with experience of dealing with all
aspects of a building's performance
Tel. 0171–240–8070

Building Research Establishment
Bucknalls Lane
Garston
Watford
Herts
WD2 7JR
Tel. 01923–894040

Building Use Studies Ltd
42–44 Newman Street
London
W1P 3PA
Consultants in the management and social science aspects of the indoor
office environment
Tel. 0171–580–8848

Carter Hodge
Solicitors
2 Liverpool Avenue
Ainsdale
Southport
Merseyside
PR8 3LX
Environment law specialists
Tel. 01704–577171

Chartered Institute of Environmental Health
Chadwick Court
15 Hatfields
London
SE1 8BJ
Tel. 0171–928–6006

Chartered Institution of Building Services Engineers
222 Balham High Road
London
SW12 9BS
Tel. 0181–675–5211

Dr Keith Eaton
Consultant Physician
Templars
Terrace Road South
Binfield

Berks
RG42 4DN
Medical specialist in allergies
Tel. 01344–53919

Health and Safety Executive
St Hugh's House
Stanley Precinct
Bootle
Merseyside
L20 3QY
Tel. 0151–951–4000

Kinsley Lord
34 Old Queen Street
London
SW1H 9HP
Tel. 0171–222–7122

National Institute of Public Heath
Sroborova 48
10042
Prague
Czech Republic
Tel. +42–267310283

Procord Ltd
6th Floor
Baltic House
Kingston Crescent
Portsmouth
PO2 8QL
Tel. 01705–230500

Jack Rostron MA, MRTPI, ARICS
Liverpool John Moores University,
Clarence Street
Liverpool
L3 5UG
Surveyor specializing in the diagnosis and treatment of buildings with sick building syndrome.
Tel. 0151–231–3282/01704–568432

Royal Institute of British Architects
66 Portland Place
London
W1N 4AD
Tel. 0171–580–5533

Royal Institution of Chartered Surveyors
12 St George Street
Parliament Square
London
SW1P 3AD
Tel. 0171–222–7000

Weatherall, Green & Smith
Chartered Surveyors
Norfolk House
31 St James's Square
London
SW1Y 4JR
Major firm of chartered surveyors with experience of commercial office
buildings
Tel. 0171–493–5566

World Health Organization
Regional Office for Europe
8 Scherfigsvej
DK 2100
Copenhagen
Denmark
Tel. 010453–9171717

Appendix B:
Expert systems

Nick McCallen

The use of computational techniques in the context of SBS may be divided into three broad areas:

- identification of SBS as an existing or potential problem;
- analysis leading to identification of possible causal factors;
- generation of recommendations for remedial action.

Some would argue that the existence of SBS is clearly demonstrable without the need to resort to computers, and that adherence to the various building and maintenance standards should eliminate the potential for SBS. There may be some validity in these propositions, but their pursuance would tend to isolate identification from the diagnostic and remedial processes.

As the Building Research Establishment (BRE) has stated [1]: 'SBS begins to look like an effect without a cause: no single factor has been identified as responsible for SBS, and there are reasons to doubt each suggested explanation.'

To isolate identification of the existence of SBS in a building would therefore be less than constructive, as it is not possible to say 'SBS exists, therefore we must do this. …' Even less is it possible, or even sensible, to say 'If we do this, this and this, our building will not suffer from SBS!

Clearly a more constructive approach is to identify as many of the indicators of SBS as possible and pursue a structured investigation of their contributory factors. Such an investigation inevitably involves significant record-keeping and crosschecking – the very things at which computers excel, even in the hands of non-specialists. Use of appropriate software in the first area consequently facilitates the maximization of information gleaned from the identification stage, to be passed to the second, analytical, stage.

Much important work has been carried out in the development of specialized computational aids to the analysis of possible causal factors.

These include advanced modelling techniques for investigating specific areas such as air flow within a defined space [2].

Most of these advanced techniques require a considerable amount of specialized expertise in computing, and their detailed treatment is consequently beyond the scope of this book.

These techniques will produce suggested solutions to specific problems, and in many cases allow the testing of solutions by modelling before expensive installation is undertaken. Unfortunately, solutions produced by these techniques tend to treat individual symptoms, which may not actually produce a cure for the overall problem. They are certainly of great value in assessing specific causal factors, but must not be seen as global solutions.

One type of software application which requires little, if any, specialist knowledge of computing, but which addresses all three of the areas identified at the head of this appendix, is the expert system.

An expert system is a piece of software which aims to mimic the logic processes utilized by a human expert in submitting received information to the expert's accumulated knowledge (the knowledge base). The software applies a set of rules to the processing of the input information in order to reach the same conclusion, as the human expert would, given the same initial information. A potential advantage of the software system over the human expert is the ability to combine knowledge bases acquired from several experts, creating an environment in which differing areas of expertise may interact in a single consultation, producing a single solution or set of recommendations, without the need for multiple, very expensive, consultations with a number of different experts in circumstances where simple actions may provide the solution to many problems.

Other important features of the expert system include:

- the ability to add to the knowledge base, without any reprogramming being required;
- the ability to automatically generate an explanation for recommendations;
- the ability to accept information in virtually any order and still process it consistently;
- the ability to represent and provide access to a large body of information in a user-friendly manner.

User-friendliness is rated as being very important in an expert system, as the system is intended for use in a context of non-computer-related problems, so the user has every right to expect communication with the software to be simple so as not to create another layer to the problem to be solved.

At the time of writing, one such expert system specifically designed to be used in the context of SBS was SBARS (the Sick Building Assessment and Rehabilitation System) [3]. The remainder of this appendix will consider SBARS as a case study in the use of expert systems in the investigation of SBS.

The purpose of SBARS

SBARS aims to identify causes of SBS and to recommend actions to rectify it. It also evaluates generally the quality of the indoor environment and presents solutions to deficiencies. SBARS is applicable both to existing buildings and to the design of new buildings. The system is designed for operation by a range of users, none of whom is a computer 'expert'.

The ultimate output of the system is a detailed report, broken down into up to 27 sections, containing recommendations and observations relating to the current condition of the building and steps suggested to improve the working environment within the building.

Summary of methodology

SBARS presents an initial questionnaire under three headings:

- medical;
- working environment;
- building characteristics.

Each of these headings is divided into logical sections, with a variable number of questions in each section. Any or all of the sections under any or all of the headings may be answered, completely or in part. The responses to this initial questionnaire are then subjected to a preliminary analysis which identifies aspects of the building which may give rise to problems. This preliminary report may be printed in draft form, or imported into the user's word processor for formatting and editing.

The preliminary analysis is then passed to an advanced expert system. Sections identified in the preliminary analysis as being possible sources of problems may then be investigated further by a sequence of more detailed questions determined by both the initial responses and by the responses to the detailed questions themselves. A detailed report results from this process, and may be displayed on screen, printed in draft form, and/or saved to file for importing into a word processor.

The knowledge base

The knowledge base utilized by SBARS has, as indicated in the previous section, three contributory subsystems:

1. medical – dealing with complaints from occupants arising from physical symptoms experienced;

2. working environment – dealing with occupants' complaints about working conditions in the building under investigation;
3. building characteristics – dealing with the structural and operational aspects of the building. It is these aspects which contribute to the over-all quality of the working environment in the building, so naturally they are subject to the greatest detail in questioning.

While each section is initially investigated with its own preliminary questionnaire, the responses received are cross-referenced, so that all potential problem areas are flagged no matter which section is being answered. Figure B.1 illustrates the relationships between the three sections of the knowledge base and a set of markers (flags) used to indicate possible problem areas in the structure, maintenance or management of the building (referred to as building characteristics).

The initial questionnaires and report

In many senses, computer software can be compared to a television set in that the user does not need to know how it works in order to use it effectively. This philosophy is embraced by all modern software, providing a simple interface for the user. Any system like SBARS which is designed for a range of users must also attempt to ensure that the user knows exactly what input is expected, and that that input conveys all necessary information to the system. SBARS keeps data entry operations simple, while ensuring that the kind of information required by the system is provided.

All questions in the preliminary questionnaires allow just three responses – Y(es), N(o), or ?(Don't know). As a safety feature, any 'Don't know' responses are treated as a potential 'trouble flag'. The questions are phrased so as to vary the significance of Y and N. This method of response collection has a double purpose in that it makes input simple, while ensuring that only sensible and useable data is entered.

The three questionnaires each consider responses in a number of areas, listed in Table B.1, which may contribute to or indicate the existence of SBS. Each questionnaire pursues a line of questioning appropriate to its

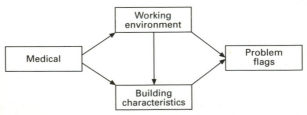

Fig. B.1 Relationships between components of the knowledge base.

Table B.1 Areas of preliminary questionnaires

Building characteristics	Medical factors	Working environment
Ventilation	Eye symptoms	Air movement/quality
Humidity	Nasal symptoms	Humidity/temperature
Heating	Throat symptoms	Noise discomfort
Furnishings/colour	Breathing problems	Lighting complaints
Maintenance	Skin problems	Construction of building
Cleaning	Aches and pains	Use of building
Lighting	Stress symptoms	Management of building
Building management	Neurotoxic symptoms	Furnishing
Use of building	Odour/taste complaints	Maintenance
Contaminants	General complaints	Cleaning
Noise management		

own area, and is able to flag any building characteristic if the users' responses suggest that area to be suspect.

There is a facility for printing the entire initial questionnaire, creating a useful paper checklist for maintenance purposes. A section of the print-out of the initial questionnaire appears as Table B.2. The full printed questionnaire runs to more than 20 pages, and is intended to be used either as a record for collecting responses 'in the field' to be entered into a computer on return to base, or as a free standing tool for checking procedures etc.

The medical questionnaire concentrates on recording the frequency of complaints related to a range of medical symptoms which may be experienced by building occupants. From this information, particular building characteristics, such as the nature or design of the ventilation system may be flagged as suspect. The working environment questionnaire adopts a similar approach, considering comments from building occupants relating to their satisfaction or otherwise with a number of factors such as levels of lighting. Again, areas of building characteristics may be flagged for further investigation. The building characteristics questionnaire, as the name implies, deals directly with the features of the building likely to be concerned in any incidence of SBS.

Thus the system allows for three distinct lines of investigation, or methods of use, reflecting the potential of the expert system to be used as a check during design, as a tool in routine maintenance and management, or as a diagnostic tool in response to environmental or medical complaints.

At the request of the user, the preliminary responses are submitted for initial analysis, which identifies areas which may contribute to SBS. These areas are flagged for further investigation at the next stage of the system, as mentioned earlier.

Table B.2 An extract from the paper version of the initial questionnaire

---

Section 6 – Cleaning

Circle Y or N to show responses:

Is the building fabric regularly cleaned; including exterior windows?	Y	N
Are internal surfaces including carpets, floors and furniture regularly cleaned?	Y	N
Does regular dampdusting take place on all hard surfaces?	Y	N
Are cleaning fluids and chemicals used correctly to manufacturer's specification?	Y	N
Is the cleaning plant used as per manufacturer's specification?	Y	N
Are air vents and diffusers regularly cleaned?	Y	N
Are luminaires regularly cleaned?	Y	N
Are air filters cleaned as per manufacturer's specification?	Y	N
Are ventilation ducts inspected and cleaned as necessary?	Y	N
Are heating/cooling coils regularly cleaned and inspected?	Y	N
Are the insides of filing cabinets regularly vacuumed?	Y	N

At the end of this process, the user has a choice of

1. creating a preliminary report file on disk for later printing or import into a word processor for customization. Files created here are stored with the name selected for the preliminary response file, but with the '.TXT' extension added automatically. These are standard DOS TEXT files;
2. printing a copy of the preliminary report;
3. returning direct to the main system for processing of the results of the preliminary analysis, without creating any output from this phase.

Selected sections of a preliminary report appear as Table B.3.

Detailed analysis

Selecting detailed analysis calls up a screen listing all the sections of the three questionnaires. A number of the selection brackets for each section may be shaded in, indicating that these section have been found to be clear of problems on the basis of the user's responses to the initial questionnaires.

Upon selecting a section for detailed analysis, the user is presented with a sequence of questions determined by (1) responses to the preliminary questionnaires; (2) the analysis performed by the system; and (3) the users response to earlier questions in detailed analysis.

Table B.3 Sections of a preliminary report

Section 1 – Ventilation
Your responses indicate potential problems in this area. 68 responses to questions related to this section were considered, of which 18 (26%) indicated this area as a possible source of problems.

Section 10 – Contaminants
Your responses indicate potential problems in this area. 54 responses to questions related to this section were considered, of which 17 (31%) indicated this area as a possible source of problems.

Section 15 – Breathing problems
Your responses indicate possible problems in one or more of the following areas:

Ventilation
Contaminants
Heating
Humidity

Each question provides four options, selected by the user pressing a number key to correspond to a choice on the mini menu presented with each question.

Choice 4, 'unknown', should be used if the user cannot answer 1 'yes' or 2 'no' to a question for one of a variety of reasons. This is usually taken as a possible problem indicator.

Option 3, 'Why?', calls up a brief explanation of the thinking behind the question, and in many cases some information to help the user decide on the appropriate answer. The question is repeated immediately after the explanation. Table B.4 presents an example of such an explanation.

Table B.4 An example of an explanation requested for a question in detailed analysis

Heating Factors
2. Was the heating system designed to achieve space temperatures between 19–21°C in winter heating conditions?
    1) yes
    2) no
    3) why?
    4) unknown

3.
    CIBSE Guide A recommends design conditions of 20–22°C for a space. Temperatures will vary upwards of this figure, however. Research has shown that temperatures between 19–21°C appear more comfortable as there is less fatigue at cooler temperatures. Also warm air appears stuffy and possibly polluted.

At the end of each section's questioning, the system presents the recommendations for that section on screen, pausing for a keypress between each one until the end of the recommendations for the current section is reached, and then returns to the section selection screen. If the print option is turned on, the recommendations for that section are sent to the printer during this process. If the file output is turned on, a consultation record file is updated. The user may now select another section for analysis.

The ultimate objective is, naturally, to produce a report detailing recommended actions. The facility to create a consultation file provides this. The file is in plain text format, so may be imported into virtually any word processor for embellishment and/or incorporation into other documents. This report, typically 10–12 pages long, may be used in many ways, from an action-plan for in-house maintenance to the basis of a consultant's report. Table B.5 shows an extract from a sample report as produced by the system.

REVIEW AND CONCLUSIONS TO BE DRAWN FROM THE CASE STUDY

Expert systems such as SBARS are not intended to replace specialists in the areas covered by the systems – indeed, without the specialists the expert systems could not be created. Rather they may be used as a preliminary to specialist consultations, and in many cases are used by the specialists themselves. SBARS will produce a detailed set of recommendations, but in certain cases will refer the the user to specialists in specific fields, rather like a general practitioner referring a patient to a neurologist.

Any kind of expert system depends on the knowledge of the experts used in creating the knowledge base. As mentioned previously, SBARS works under three principal headings, but does not pretend to represent the highly specialized skills and current knowledge of, for instance, the heating engineer. Rather, SBARS uses rules derived from fairly well-fixed rules in the three areas referred to. This is not to say that the knowledge base of SBARS is immutable. It is clearly intended that the knowledge base employed by SBARS will be constantly updated and expanded as knowledge of SBS evolves.

SBARS will seek to identify, from the information provided by the user, all factors possibly significant to SBS relating to the building under study, and will generate a report recommending many actions. SBS is seen to have a huge range of causal factors, with unerring identification and eradication of those factors presenting enormous difficulty, as correction of one suspected cause may perhaps create another. The most practical way to establish procedures with reasonable chance of success in eliminating

Table B.5 One section from a final report

Ventilation recommendations

Consult system designer about ventilation rates achieved by the a/c/ventilation system in relation to those specified in CIBSE guides. These rates are defined as being the minimum required to ensure correct ventilation.

The objective of introducing fresh air into a space is to dilute the level of contamination to one which is safe and acceptable, and may be achieved by ensuring that the levels of ventilation specified in the CIBSE guides are attained.

Increase ventilation rate to speed removal of tobacco smoke and see recommendations in other sections.

Contact an environmental consultancy for test of carbon dioxide levels in the space, increase ventilation rate and see recommendations in other sections.

Increase ventilation rates to achieve the recommended fresh air rates in CIBSE Guide B. They are

(a) 8 Lt/sec.   (minimum)
(b) 16 Lt/sec.   (where some smoking)
(c) 24 Lt/sec.   (where heavy smoking)

Also see recommendations in other sections.

If the prevailing wind does not disperse potential sources of pollution, there is more than a possibility of introducing contaminated air into the building.

Review siting of air intake vents relative to the prevailing wind and potential sources of pollution.

Determine the direction of the prevailing wind from the local meteorological office to assess the risk of contaminated air being introduced into the building.

Check and if necessary replace air filters with those specified by the manufacturer of the system.

Ensure housings and seals are checked for damage before fitting new air filters.

Most modern window units have inbuilt vents to allow ventilation. This gives staff the ability to ventilate as they require, and also provides a method of ventilation in the case of plant failure.

At next window replacement, consider openable, lockable windows with vents.

Remove items of furniture obstructing diffusers.

Remove files, records, etc. from vicinity of diffusers.

Consult a/c/ventilation system designer as to the positioning of the vents.

Consider the practicality of installing exhaust luminaires.

SBS, or avoiding its occurrence, seems to be the concentration of knowledge enabled by software expert systems.

In allowing analysis of SBS-related problems by means of a choice of approaches, SBARS typifies the flexibility achievable by computer software. As the knowledge base is expanded, incorporating expertise from an ever widening range of specializations, the accuracy and specificity of diagnosis may be expected to increase, as will the number of ways of applying the system.

REFERENCES

1. Raw, G. (1990) *Sick Building Syndrome*, Building Research Establishment, Watford.
2. Hedges, A. (1996) *Computational Techniques concerning Sick Building Syndrome*, Cornel University, Internet.
3. Environmental Management & Intelligent Systems Ltd., EMIIS Ltd (1994) *Sick Building Assessment and Rehabilitation System*, v.2.1, Salford.

Enquiries about the full working version of S-BarS should be made to:

S-BarS
c/o Michael Doggwiler
E & FN Spon
2–6 Boundary Row
London
SE1 8HN
UK

tel +44 (0)171 865 0066
fax +44 (0)171 522 9623

Index